Surface Preparation Techniques for Adhesive Bonding

Surface Preparation Techniques for Adhesive Bonding

Second Edition

Raymond F. Wegman and
James Van Twisk

AMSTERDAM • BOSTON • HEIDELBERG • LONDON • NEW YORK • OXFORD
PARIS • SAN DIEGO • SAN FRANCISCO • SINGAPORE • SYDNEY • TOKYO
William Andrew is an imprint of Elsevier

William Andrew is an imprint of Elsevier
The Boulevard, Langford Lane, Kidlington, Oxford OX5 1GB, UK
225 Wyman Street, Waltham, MA 02451, USA

British Library Cataloguing-in-Publication Data
A catalogue record for this book is available from the British Library

Library of Congress Cataloging-in-Publication Data
A catalog record for this book is available from the Library of Congress

ISBN: 978-1-4557-3126-8

For information on all Elsevier publications
visit our web site at books.elsevier.com

Printed and bound in the US

12 13 14 15 16 10 9 8 7 6 5 4 3 2 1

Dedication

To all those engineers and scientists who have dedicated their careers to the advancement of the field of materials and processing for adhesive bonding; and to my wife, Rose, whose encouragement, indulgence, and sacrifice made this book possible.

Contents

Preface

The purpose of this handbook is to provide information on processing adherends prior to adhesive bonding. Where sufficient data were available the processes are given in the form of process specifications. Further, where available, data are given to provide potential users with a basis for the selection of the process most suitable for their particular application and facility.

It is known that many of the chemicals used in these processes are hazardous due to their strong oxidizing properties, or may be toxic or hazardous to one's health. It is the user's responsibility to assure that proper safety, handling and disposal procedures are implemented and monitored when any of these methods is employed.

Raymond F. Wegman

Acknowledgments

My grateful thanks to Dr. David W. Levi for his help and encouragement; Dr. John D. Venables and his staff at the Materials and Surface Science Department at Martin Marietta Laboratories, for supplying the numerous micrographs and drawings of treated surfaces.

Raymond F. Wegman

Notice

To the best of the Publisher's knowledge the information contained in this publication is accurate; however, the Publisher assumes no liability for errors or any consequences arising from the use of the information contained herein. Final determination of the suitability of any information, procedure, or product for use contemplated by any user, and the manner of that use, is the sole responsibility of the user.

Mention of trade names or commercial products does not constitute endorsement or recommendation for use by the Publisher.

The book is intended for information only. It is known that many of the chemicals used in the processes described are hazardous, due to their strong oxidizing properties, or may be toxic or hazardous to one's health. It is the user's responsibility to assure that proper safety, handling and disposal procedures are implemented and monitored when any of these methods is employed. Expert advice should be obtained at all times when implementation is being considered.

1 Introduction

1.1 Adhesion

Adhesion is a surface phenomenon, i.e., adhesion is controlled by the condition of the surface of the adherend. The ASTM [1] defines adhesion as "the state in which two surfaces are held together by interfacial forces which consist of valence forces or interlocking actions or both." Adhesion between surfaces which are held together by valence forces is called specific adhesion; this is the same type of force which gives rise to cohesion. Cohesion is defined as the state in which particles of a single substance are held together by primary or secondary forces. As used in the field, cohesion is defined as the state in which the particles of the adhesive (or the adherend) are held together.

Adhesion between surfaces in which the adhesive holds the parts together by interlocking action is known as mechanical adhesion.

Both specific adhesion and mechanical adhesion are important to the understanding of how adhesion is affected by surface preparation. Allen [2] in his discussion of the fundamentals of adhesion concluded by stating "... an adhesive bond achieves its strength from the combination of a variety of sources; (mechanisms)... For these mechanisms, the relative importance and the proper way which they should be combined will vary from one example to another, but none should be excluded without very careful consideration and exploration."

1.1.1 *Specific Adhesion*

According to the theory relating to specific adhesion, this type of adhesion involves the establishment of some kind of attraction between the atoms and the molecules which make up the adhesive and the adherends. These attractions may involve primary bonding forces, which tend to be quite strong; hydrogen bonding, which yields intermediate strength; and the weaker secondary (Van der Waals) forces.

Primary bonding may be covalent or ionic in nature. Covalent bonding involves the sharing of electron pairs between adjacent atoms. Ionic or electrostatic forces are the type of primary bonds that are found in ionic crystals. Another type of primary bond is the metallic bond which is similar to the covalent bond except that it involves the valence electrons in the metal. This type of bonding is discussed by Verink [3], Wegman and Levi [4] and Salomon [5].

Surface Preparation Techniques for Adhesive Bonding.

Secondary bonding involves dipole-dipole interactions, induced dipole interactions and dispersion forces. Secondary bonding becomes important in adhesion when nonpolar or chemically inert surfaces are involved. Hydrogen bonding may be considered a special case of dipole interaction, since hydrogen bonds result from the sharing of a proton by two electron-negative atoms. However, hydrogen bond strengths are of the same order as a weak primary bond.

Another type of bond, discussed by DeLollis [6], is described as a chemisorbed bond. This type of bonding is proposed as the reason why adhesive promoters, such as primers and coupling agents, are successful in overcoming potentially weak boundary layers and result in good durable bonds.

In dealing with specific adhesion, good contact must be obtained between the adhesive and the surface of the adherend. To obtain this molecular contact with a solid adherend requires the wetting of the solid surface by the liquid adhesive. During the bonding process even the solid adhesive must go through a liquid phase. To understand the conditions of adequate wetting one must consider the role of surface energetics in adhesion. This, simply stated, requires that in order for a liquid to wet and spread on a solid surface the critical surface tension of the solid or solids must be greater than the surface tension of the liquid. In the case of polar solids such as metals and metal oxides this requirement is easily met, because the surface energies of the solid, provided that the surfaces are clean, are greater than 500 dyne/cm while the surface energies of the liquid adhesives are less than 500 dyne/cm [4]. However, even if good wetting and good contact between the adhesive and adherend are obtained, it may be difficult to obtain good durable bonds if there is a weak boundary layer on the adherend. Therefore, it is necessary to carefully select the proper surface preparation technique for the particular adherend, in order to make sure that such a weak boundary layer does not occur, which would make the bond practically useless. Surface energetics is further discussed by Kaelble [7].

1.1.2 Mechanical Adhesion

Mechanical adhesion results from an interlocking action between the surface structure of the adherend and the adhesive or primer.

Bickerman [8] proposed that adhesion was due to the inherent roughness of all surfaces. He accepted the fact that molecular forces of attraction caused an adhesive to wet and spread on the surface. Once this was achieved, however, Bickerman felt that mechanical coupling between the adhesive and the inherently rough adherend was more than enough to account for bond strength. Surface roughness can also account for a number of negative factors such as trapped gas bubbles, as described by DeBruyne [9], and imperfect molecular fit, as described by Eley [10]. Many arguments have been presented against Bickerman's theory and inherent surface roughness should be considered to be a contributing factor only, rather than a basic factor in the theory of adhesion.

In 1977, Chen et al. [11] proposed that the surface of Forest Products Laboratory (FPL) etched aluminum consisted of fine finger-like structures of aluminum oxide which were determined to be about 50 Angstroms thick and approximately 400 Angstroms in height. Each of the fine fingers protruded from the tripoint of the oxide structure. This finding was made possible by advanced electron microscopy using such techniques as the scanning transmission microscope. Similar findings were made when the surface of phosphoric acid anodized aluminum was studied.

Further discussions and stereomicrographs of treated aluminum surfaces are presented in Chapter 2 of this book.

In 1980, Ditcheck et al. [12] described the morphology and composition of titanium adherends prepared for adhesive bonding. In this work they discussed macro-rough and micro-rough surfaces. From their determination of the morphology of various surfaces they predicted the order of bondability of the surfaces and the reliability of the resultant bonds. Their predictions were confirmed by Brown [13] using the wedge test and by Wagman and Levi [14] using stress durability testing. This work is described in more detail in Chapter 3. Some excellent micrographs of these surfaces were presented by Venables [15] using extended resolution scanning electron microscopy.

This work shows that mechanical adhesion on a microscopic level does play a major role in adhesion and the durability of the resultant bonds. Therefore, as in the case of specific adhesion, it is equally important that proper surface preparation be used when one considers mechanical adhesion.

Whether one subscribes to the specific adhesion theory, the mechanical adhesion theory, or a combination of the two, surface preparation is a major item that must be considered.

At a National Research Council Workshop held in Washington DC [16] it was suggested that any study of the interface should begin with the realization that the traditional picture of the two-dimensional nature of the interface is outdated. It was further suggested that the interface is a three-dimensional "interphase." This interphase was described as "extending from some point in the adherend where the local properties begin to change from the bulk properties, through the interface, and into the adhesive to where the local properties approach the bulk properties." The interphase can extend from a few to a few hundred nanometers. On the adherend side it includes morphological alterations in the adherend near the surface, as well as oxides, whether deliberately constructed or native to the adherend surface. The oxide layer can vary in porosity and microstructure. Absorbed gases may be present on the oxide surface giving rise to conditions unfavorable to a good durable bond. A polymeric adhesive may be affected by the presence of the adherend surface. The polymeric network may be altered in the region close to the adherend surface, giving rise to a material which is different in composition and mechanical properties. This is caused by diffusion of certain adhesive components into the oxide surface leaving a layer of adhesive which contains a smaller amount of these components than does the bulk. Acceptance of the interphase as a working model of the actual joint interface leads to the conclusion that the logical beginning of a strong, durable adhesive bonded joint is the selection of the proper surface treatment.

1.2 Bonding

Adhesive bonding is widely accepted as the method for joining material, distributing stress, and producing structures that could not otherwise be made. The successful performance of these structures depends upon the ability to control all of the independent steps of the production line.

Wegman and Levi [17] describe the various considerations involved in setting up a satisfactory adhesive bonding shop. A summary of these steps includes:

1. Receiving materials. These materials include adherends, adhesives, and the chemicals and solvents required for cleaning.
2. Knowledge of the fabrication processes for the parts to be bonded. The process by which the individual parts were produced may affect the type of cleaning process which will be required. The use of cutting oils, release agents, or the type of binder used in the manufacture of a casting mold can have an effect upon the selection of a cleaning process and the success of subsequent bonding steps.
3. Mating of parts to be bonded. Some cleaning processes such as acid etching will remove material, while processes such as anodizing and other coating processes will add material. Therefore, in some designs it is important to know how much material was removed or added to allow for a satisfactory glue line. Machined parts should be mated prior to application of adhesive to insure there are no misfits which could cause voids, surface bulges or stress concentration points.
4. Surface preparation. This is often the most critical, complex, and frequently abused step in the bonding sequence.

It is important to remember these factors regarding surface preparation:

1. Know the adherend material, its surface condition and how it was prepared.
2. Use good safety and health practices when using cleaning agents.
3. For wiping parts, only clean white cotton cloths with no sizing or finish or paper towels with little or no organic binders should be used. Never use synthetic materials.
4. Hot solvent washing or vapor degreasing should not be used to clean composite materials or for secondary bonding preparations, unless these materials have been tested and proven to be safe.
5. Train, qualify and requalify personnel who operate the cleaning facility.
6. Check the cleaning line regularly.
7. Cleanliness, vigilance and attention to detail are of primary importance on the job.
8. Don't rely on faith. Know your process, materials, personnel and quality control tests and results at all times.

Another important but often neglected step in processing for adhesive bonding is quality control. Control of the temperature of the solution, control of the drying between immersions and control of the type and flow of rinse waters are all very important. One of the most critical steps requiring control is the rinsing of adherends. McNamara et al. [18] concluded that "the rinsing operation during FPL pretreatment of aluminum surfaces for bonding is crucial to the success of the bonding operation." Immersion rinse solutions with a $pH < 3.0$, if allowed to dry on the adherend surface, can lead to catastrophic bond failure. This control of rinsing is required for other treatments as well. McNamara et al. [19] also pointed out that aging studies on standard and optimized FPL solutions indicate that no significant change in oxide morphology occurs up to very high levels of dissolved aluminum (approximately 2 wt % Al). Long before this level is reached, reaction by-products precipitate from the solution as it cools to room temperature. This may cause rinsing problems. The authors suggest that a specific gravity measurement be used to monitor the useful life of the tank. Shearer [20] recommends the monitoring of wedge test [21] results as the best control of the etch tank. The floating roller peel test [22] can also be used as a quality assurance test to monitor the etch tank [20,23].

Shearer [19] also states that higher than normal etch bath temperature may affect the effectiveness of an etch solution. Other points that should be considered include the health hazards involved, the protection of materials that have been prepared for bonding, the inspection of bonded parts and the development of the process specification.

1.3 Plasma Surface Treatment of Material for Improved Adhesive Bonding

Surface preparation of many of the materials in the preceding chapters to improve adhesive bonding and bond durability has required the use of strong, dangerous, and often hazardous chemicals. A newer surface preparation technology for cleaning surfaces for bonding is atmospheric plasma technology, a technology similar to corona treatment.

"Corona treatment (sometimes referred to as air plasma) is a surface modification technique that uses a low temperature corona discharge plasma to impart changes to the properties of the surface. The corona plasma is generated by the application of a high voltage to sharp electrode tips which form plasma at the ends of the sharp tips" [24]. Corona treatments are limited largely to flat thin surfaces. While the treatment speed can be high, the surface activation is marginal (usually up to 60 dyne/cm). Corona does not remove organic contamination from the surface, exposes an electrical charge, and produces large amounts of ozone which are detrimental to human health and very corrosive to metals.

Another more effective treatment is achieved via atmospheric pressure (air) plasma (APP). "Like corona, atmospheric pressure plasma is the electrical ionization of a gas. The plasma (glow) discharge creates a smooth, undifferentiated cloud of ionized gas with no visible electrical filaments. Unlike corona, plasma is created at a much lower voltage" [25]. APP treatment is one of the most efficient surface treatment processes for cleaning, activating, or coating materials like plastic, metal (such as aluminum and titanium), or glass. APP improves low surface energy levels (22–32 dyne/cm) to as high as achieving complete wetability (72 dyne/cm). High dyne levels generally associated with better adhesion for various adhesives, paints, inks, and coatings. APP activates surfaces by creating functional groups such as carbonyl and hydroxyl groups (with polymer substrates). Atmospheric plasma is designed for industrial, fast throughput applications. The length of treatment effects can last for a long time (if treated parts are properly stored and contamination is avoided), but typically treatment and subsequent processes occur in back-to-back steps.

"Plasma surface treatments allow for almost limitless surface modifications. Materials can have any size, from nano-sized components to endless sheets in continuous production. A wide range of materials can be treated with plasma surface technology, including glass, metals, metalloids, rubbers, and polymers" [26]. Atmospheric plasma devices (jets) can be incorporated into automated work cell solutions effectively treating both 2D and 3D surfaces at high speeds.

During cleaning, atmospheric plasma removes molecules deposited on the surface such as greases, oxide layers, or even silicone rendering an ultraclean surface for adhesive bonding. Advantages of using plasma treatment include no grit or media are generated that require removal or containment, no harsh or hazardous chemicals are required, resulting in safer work environments and the process parameters can be precisely controlled.

Some important industrial applications include the treatment of sandwich panel component for the manufacture of refrigerated trucks using Openair®Plasma [27–29], primer-free bonding of windshields [30], battery technology, solar thin film cell manufacturing, printed circuit boards, metal cleaning, and bonding to foams [31].

A large variety of substrates can be treated by atmospheric plasma devices but require specialized equipment solutions. The selection and application methods have vastly grown and improved in the last decade. A few equipment suppliers have emerged that can be contacted for help in selection of the best equipment and conditions for the particular materials, configuration, application, and environment that will be involved.

In 1995, a company, Plasmatreat GmbH based in Steinhagan, Germany, invented a new technology called Openair®Plasma (or APP). Some of the advantages of this technology include high process reliability, high cost effectiveness, high degree of activation, large process window, easy to integrate, and environmentally friendly [32]. The requirements to run a plasma pretreatment system are public current (230 V/400 V) as well as oil-free compressed air and an extraction system for health and safety. No ozone is released but nitrogen oxides and monoxides may be produced when certain materials are pretreated. The only operational expenses will result from the electricity and compressed air consumed. The plasma treatment can be used on complicated shapes [33].

Indications are that all of the materials covered in Chapters 2–9 could be surfaces treated for adhesive bonding by plasma technology. In 2009, Wesley Taylor published a "Technical Synopsis of Plasma Surface Treatments." In his conclusions, he stated "all of the different plasma treatment systems have their specific benefits. Air plasma can treat most of the types of plastics as well as some other materials with few limitations. Atmospheric chemical plasma is an incredible breakthrough because it removes any previous limitations created by previous methods of surface treatment" [34].

References

[1] ASTM D907, Standard definitions of terms relating to adhesives, Annual Book of ASTM Standards vol. 15.06.
[2] K.W. Allen, Fundamentals of Adhesion—The Science Beneath the Practice, Proceedings, Fifth International Joint Military/Government-Industry Symposium on Structural Adhesive Bonding at U.S. Army Armament Research, Development and Engineering Center, Picatinny Arsenal, Dover, NJ, 3–5 November, 1987, pp. 1–22.
[3] E.D. Verink Jr., Structure of matter and introduction to metallurgy, chapter 2. Part A in: Young, Shane (Eds.), Materials and Processes, third ed., Marcel Dekker, Inc., New York, NY, 1985, p. 672.
[4] R.F. Wegman, D.W. Levi, Adhesives, chapter 12. Part B in: Young, Shane (Eds.), Materials and Processes, third ed., Marcel Dekker, Inc., New York, NY, 1985, pp. 671–685
[5] G. Salomon, Chapter 1, in: Houwink, Salomon (Eds.), Adhesion and Adhesives, Elsevier, New York, NY, 1965.
[6] N.J. DeLollis, Adhesion Theory, Review and Critique, SS-RR-70-915, Sandia Laboratories, December, 1970.
[7] D.H. Kaelble, Surface Energetic for Bonding and Fracture, 7th National SAMPE Technical Conference, SAMPE, Covina, CA, 14–16 October, 1975, pp. 1–9.
[8] J.J. Bickerman, The fundamentals of tackiness and adhesion, J. Colloid. 2 (1947) 174.
[9] N.A. DeBruyne, The Extent of Contact Between Glue and Adherend, Bulletin No. 168 December, 1956, The Technical Service Department, Aero Research Ltd. Duxford, Cambridge, England.
[10] D.D. Eley, Surface chemistry studies in relation to adhesion, Kolloid-Zeitschrift and Zeitschrift Fur Polymers, Band 197, Heft 1–2 (1964) pp. 129–134.
[11] J.M. Chen, T.S. Sun, J.D. Venables, R. Hopping, Effect of Fluoride Contamination on the Microstructure and Bondability of Aluminum Surfaces, 22nd National SAMPE Symposium and Exhibition, 1977, pp. 25–46.
[12] B.M. Ditcheck, K.R. Breen, T.S. Sun, J.D. Venables, Morphology and Composition of Titanium Adherends Prepared for Adhesive Bonding, 25th National SAMPE Symposium and Exhibition, 6–8 May, 1980, pp. 13–24.
[13] S.R. Brown, An Evaluation of Titanium Bonding Pretreatments with a Wedge Test Method, 27th National SAMPE Symposium and Exhibition, 4–6 May, 1987, pp. 363–376.
[14] R.F. Wegman, D.W. Levi, Evaluation of Titanium Prebond Treatments by Stress Durability Testing, 27th National SAMPE Symposium and Exhibition, 4–6 May, 1982, pp. 440–452.
[15] J.D. Venables, Review: adhesion and durability of metal-polymer bonds, J. of Mater. Sci. Chapman and Hall Ltd., London, 19 (1984) pp. 2431–2453.
[16] Committee Report, Reliability of Adhesive Bonds in Severe Environments, National Materials Advisory Board, National Research Council, NMAB-422, 1 December, 1984, pp. 31.
[17] R.F. Wegman, D.W. Levi, Structural adhesive processing, Part B, Chapter 22, Section V Materials and Processes, Marcel Dekker Inc., New York, NY, 1985, pp. 1177–1191.
[18] D.K. McNamara, L.J. Matienzo, J.D. Venables, J. Hattayer, S.P. Kodali, The Effect of Rinse Water pH on the Bondability of FPL-Pretreated Aluminum Surfaces, 28th National SAMPE Symposium, 12–14 April, 1983, pp. 1142–1154.

[19] D.K. McNamara, J.D. Venables, R.L. Hopping, L. Cottrell, FPL Etch Solution Lifetime, 13th National SAMPE Technical Conference, 13–15 October, 1981, pp. 666–675.
[20] R.M. Shearer, Five Year Metal Bonding with a Nonchromated Etch, 19th International SAMPE Technical Conference, 13–15 October, 1987, pp. 304–311.
[21] ASTM D3762, Standard test method for adhesive bonded surface durability of aluminum (wedge test), Annual Book of ASTM Standards vol. 15.06 Adhesives.
[22] ASTM D3167, Standard test method for floating roller peel resistance of adhesives, Annual Book of ASTM Standards vol. 15.06 Adhesives.
[23] ASTM E864, Standard practice for surface preparation of aluminum alloys to be adhesively bonded in honeycomb shelter panels, Annual Book of ASTM Standards, Building Seals and Sealants, Fire Standards; Building Construction.
[24] Wikipedia, The Free Encyclopedia, 30 April (2012).
[25] T. Gilbertson, Enercon industries, atmospheric plasma. What is it? and What's it good for? ENEWS Surface Treating Technology, 2nd Quarter (2001).
[26] Plasma surface technology, thierry corporation (USA) Thierry GmbH (Deutschland) <http://.thierry.com/Plasma-Applications-Cleaning-Activating>.
[27] High-efficiency plasma cleaning, activation and nanocoating of surfaces, <http://www.plasmatreat.com/plasma-tchnology/openair-atmospheric-plasma-technique.html>.
[28] FAQ: question and answers, <http://www.plasmatreat.co.uk/faq.html>.
[29] Structural adhesive bonding for truck refrigeration trailers: secure adhesion for bonds with long-term stability, <http://www.plasmatreat.com/industrial-applications/transporation/truck_trailers_mobil_homes>.
[30] Openair®Plasma activation for reliable structural bonding in vehicle manufacturing without using environmentally damaging primers, <http://www.plasmatreat.com/industrial-applications/transporation/truck_trailers_mobil_homes>.
[31] Schmitz Cargobull AG structural adhesive bonding, <http://www.plasmatreat.com/company/partners_references/industrial = partners/schmits_cargobull.html>.
[32] The future is plasma: customized surface treatment with atmospheric pressure plasma systems, <http://www.plasmatreat.com>.
[33] Plasma treatment industry applications, Enercon industries, <http://enerconind.com/treating/library/technologu-spotlights/plasmatreatment_industry_applications>.
[34] Wesley Taylor, Technical Synopsis of Plasma Surface Treatments, December, University of Florida, Gainesville, FL, 2009.

2 Aluminum and Aluminum Alloys

2.1 Introduction

For years the industrial practice for preparing aluminum surfaces for adhesive bonding was generally the Forest Products Laboratory etch, commonly referred to as the FPL etch [1]. The use of treatments containing chromates and dichromates are no longer allowed in many countries. This would include the FPL etch, chromic acid anodize, and optimized FPL etch. These processes are being covered in this book for the historical value and the fact that they are still being used in some places under a "grandfather coverage." Since the development of the FPL etch in 1950, various other methods have been developed and proposed as improvements. Among these are the "optimized FPL etch" [2,3], the phosphoric acid anodize (PAA) [3–6], the P and P2 etchants [7] and the chromic acid anodize procedures [8].

During the twenty-five years from 1950 to 1975 very little was known about the function of the various components used to make up the solutions to treat aluminum for bonding. Surface preparation was, and to some extent still is, considered to be an art rather than a science. Industry knew how to prepare the surface but very little was known as to why and how the etchants worked.

In 1975, Bethune [2], in an attempt to understand failures in adhesive bonded structures, began to investigate the methods used to prepare the surface of aluminum and how the treatment affected the durability of the bonded structure. Bethune noted that the surface was affected by dissolving aluminum in the bath prior to use. Further, the durability of the joint was affected by both bath temperature and immersion time. It was reported to be essential that the parts were at least at a minimum temperature for a sufficient time period to receive a proper etch. This meant increasing the temperature of the bath to compensate for the cooling effect of the parts, especially if thick metal parts were to be cleaned, and to increase the immersion time to allow for the parts to come to bath temperature. Industry had known for years that a new, freshly prepared etch solution was not satisfactory for treating parts and would often save a portion of the old solution to "seed the tank." Alternatively, the first parts treated were rejected and retreated. Bethune's findings led to the "optimized" FPL etch. This involves dissolving a known amount of aluminum in the etch solution at the time of preparation, as well as increasing the treatment time and temperature. Further, Bethune noted that the time between etching and rinsing was critical. This latter finding led to the installation of spray or fog heads on the holding racks to allow the parts to be kept wet during the time between etching and rinsing.

Surface Preparation Techniques for Adhesive Bonding.

In 1977, investigators at the Martin Marietta Laboratories [9] proposed that the surface of FPL etched aluminum consisted of fine finger-like structures of aluminum oxide which were about 50 Angstroms thick and approximately 400 Angstroms high. Each protruded from the tripoints of the oxide substructure. This structure will be discussed in further detail under the discussion of the FPL and the PAA processes.

Four specific treatments for aluminum and aluminum alloys will be presented in a process specification format. Discussions of the surface chemistry, as known to date, will be presented along with available data on the strength and durability for each method. The selection of a particular method for use will be dependent upon the particular facility and the type of production as well as environmental considerations. The four treatments to be discussed will be the FPL etch, the P2 etch, the PAA and the CAA processes.

2.2 FPL ETCH

The FPL etch was developed by Forest Products Laboratory in 1950 as a method for preparing Clad 24 S-T3 (present designation 2024-T3 clad) aluminum surfaces for bonding [1]. The method was refined in 1975 and is now generally referred to as the "optimized FPL etch." In this chapter the use of the nomenclature FPL etch will refer to the optimized FPL etch version.

Much of the knowledge that is available today about the FPL etch and how it functions is the result of the work of investigators at the Martin Marietta Laboratories and Martin Marietta Aerospace Company [10–14]. These investigators pointed out that the surface of the FPL etched aluminum is aluminum oxide and not aluminum hydroxide. Aluminum hydroxide is formed by the reaction of aluminum oxide and water when the bond has failed. Figure 2.1a is a stereomicrograph of FPL etched aluminum. Figure 2.1b is a proposed structure of the oxide surface. The structure of the aluminum hydroxide after failure in water shows a very different structure. This is shown in Figure 2.2. The reaction of the FPL etch solution with the aluminum alloy [12] is given in Equation (2.1).

$$2AI + H_2SO_4 + Na_2Cr_2O_7 \rightarrow Al_2O_3 + Na_2SO_4 + Cr_2SO_4 + 4H_2O \qquad (2.1)$$

The aluminum oxide formed in Equation (2.1) is dissolved by the reaction with the sulfuric acid as shown in Equation (2.2).

$$Al_2O_3 + 3H_2SO_4 \rightarrow AI_2(SO_4)_3 + 3H_2O \qquad (2.2)$$

Most of these compounds are in the form of ions in solution. The reaction of Equation (2.1) proceeds more quickly than does the reaction of Equation (2.2), thus leaving a controlled amount of the aluminum oxide on the surface.

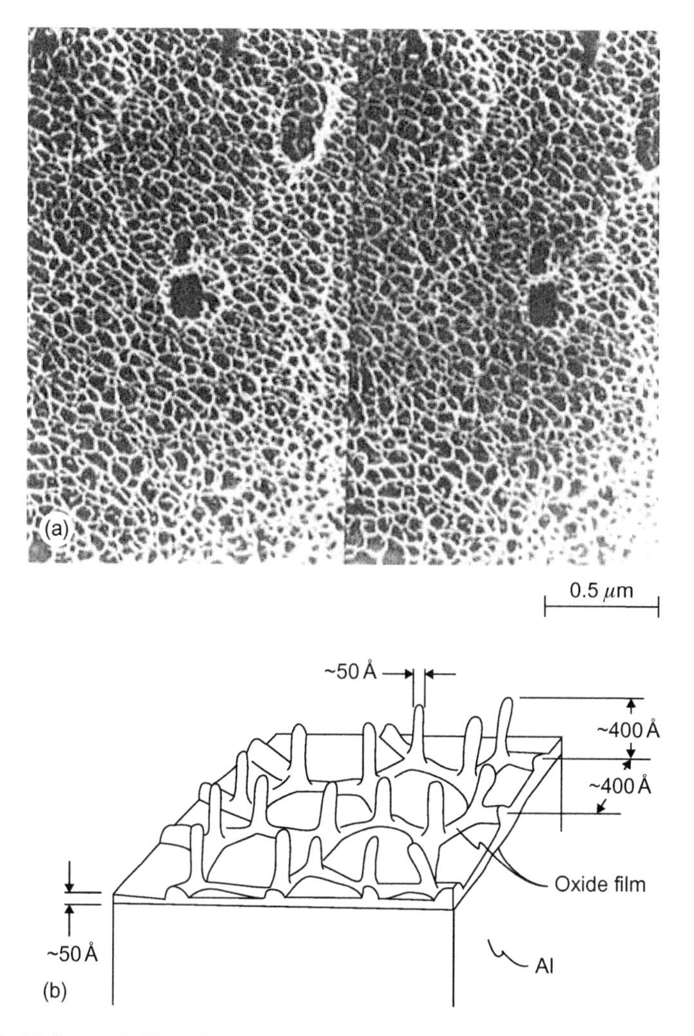

Figure 2.1 (a) Stereo STEM micrograph of oxide morphology of FPL treated 2024 aluminum surface. (b) Isometric drawing of oxide structure.
Source: Courtesy of Martin Marietta Laboratories.

One of the changes that was made to obtain the optimized etch was the addition of copper to the etch prior to use. This can be done by dissolving 2024-T3 aluminum in the etch solution [15] or by the addition of copper sulfate [13]. The copper ions in the solution appear to help form small deep pores on the surface which in turn provide a better surface for bonding.

The immersion rinse that follows the etch contains carryover solution from the parts and the racks, which will lower the pH of the rinse water. In general, for the immersion times involved in actual production, this lower pH will not have a

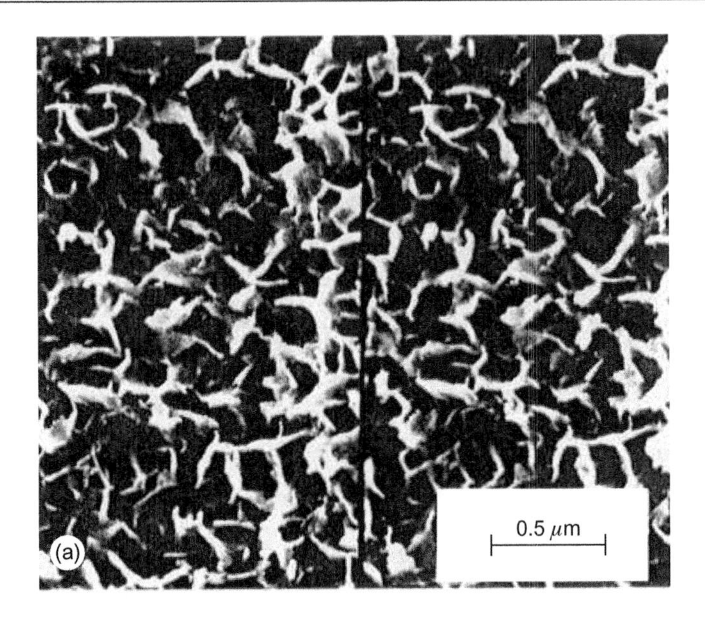

Figure 2.2 (a) Stereo XSEM micrograph. (b) Isometric drawing of aluminum hydroxide (pseudo-boehmite) produced on Al surface on exposure to moisture.
Source: Courtesy of the Martin Marietta Laboratories.

detrimental effect on the strength and durability of subsequent adhesive bonds. A final spray rinse will be effective in producing good bonding surfaces. However, immersion rinse solutions with a pH of less than 3.0 if allowed to dry on the surface may cause difficulties and can lead to bond failures [14]. Therefore, when a production facility is being designed it is recommended that a system be included, such as spray or fog heads, which will prevent the surface from drying before it is completely rinsed and ready for the next process step.

2.3 Phosphoric Acid Anodize Process

In 1975, during an examination of in-service bond failures [2] it was noted that the failures were interfacial and occurred under low stress. As a result of the extensive studies of these in-service bond failures it was determined that the weak link in the chain was the oxide layer produced by the FPL etch as it was done at that time. It was deduced that in order to develop reliable, environmentally stable bonds the oxide layer produced should be changed. As a result, the phosphoric acid anodize process known as Boeing Process Specification BAC 5555 was developed. During the years that followed many investigations have been conducted using the phosphoric acid anodize (PAA) process [16–18]. During the Primary Adhesively Bonded Structure Technology (PABST) program, conducted for the U.S. Air Force, it was felt that some of the failures that occurred during the test program may have been caused by physical damage and contamination of the oxide surface during normal prebond handling of the PAA surface.

In 1979 McNamara et al. [19]. investigated the prebond handling of the PAA surface. By the use of ultra-high resolution scanning transmission electron microscopy (STEM) they demonstrated that the surface produced by the PAA process was as shown in Figure 2.3. Basically the structure is very similar to that formed by the FPL etch except that it was about ten (10) times as thick, having a basic layer of oxide which supports a tubular cell structure. This cell structure has a thickness of about 3,000 Angstroms and supports finger-like structures which are approximately 100 Angstroms wide and 1,000 Angstroms high. The investigators concluded that the PAA process produced an oxide structure which was not easily damaged by the normal handling associated with adhesive bonding. The oxide surface shown in Figure 2.3 is formed by a two-stage process. First is the fast growth stage during which the thick pore cell structure is formed, and second is the slower stage during which the fingers or whiskers are formed on the top of the cells [20]. During this investigation it was reported that the FPL deoxidation step was not necessary for successful use of the PAA process; any deoxidation process would work. This will be discussed later.

The presence of the fluoride ion in the water used to prepare the anodize solution will affect the thickness of the oxide layer. The amount of fluoride which can be tolerated will vary with the alloy being prepared, e.g., 7075 clad will tolerate up to 1,000 ppm of fluoride ions in the PAA solution while the 2024 bare alloy will only tolerate up to 800 ppm of the fluoride ion in the solution. Therefore, the level of fluoride concentration in the PAA solution should not be allowed to exceed 600 ppm to insure a surface to which acceptable bonds can be formed [21]. Other investigations have been conducted in conjunction with use of the PAA process [22–28].

As mentioned earlier [20] it is not necessary to use the sulfuric acid-sodium dichromate etch as the deoxidation step to successfully use the PAA process. In 1981, Pocius and Claus [29] replaced the sulfuric acid-sodium dichromate step by abrasive surface conditioning. The authors used two types of Scotch-Brite

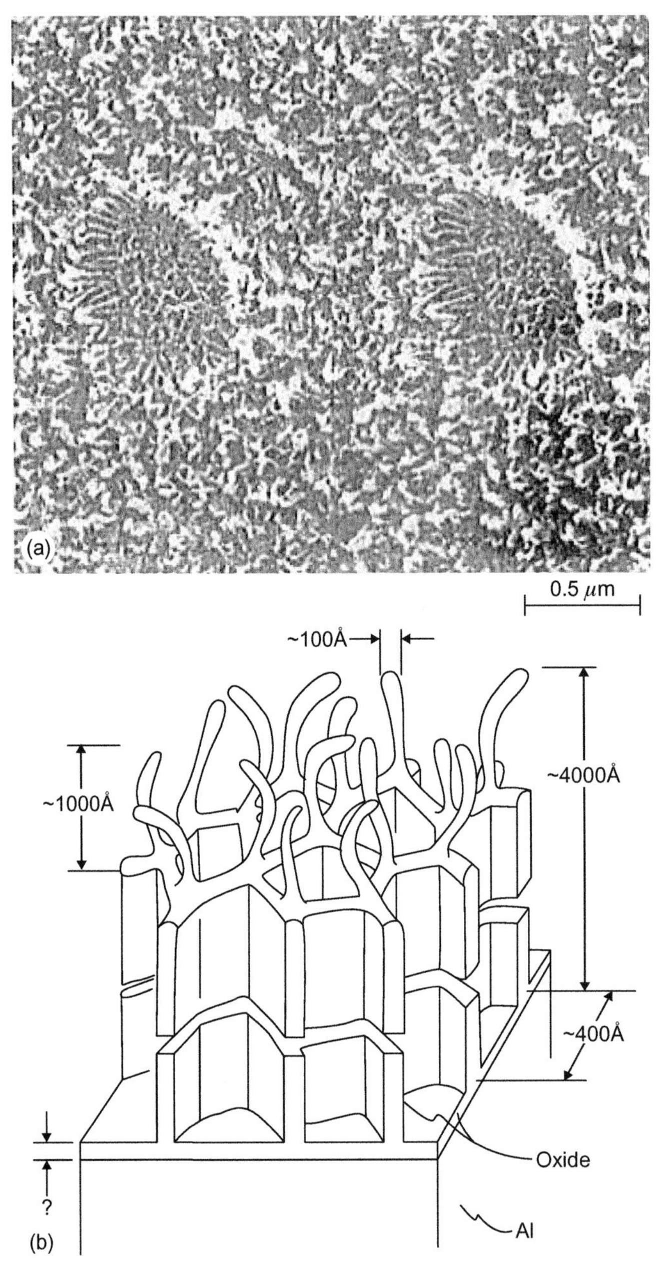

Figure 2.3 (a) Stereo STEM micrograph of oxide morphology of PAA treated 2024 Al. The depression in the center, probably due to an air bubble on the surface during anodization, allows the oxide cell walls to be seen. (b) Isometric drawing of oxide structure. *Source:* Courtesy of Martin Marietta Laboratories.

Table 2.1 Initial Physical Properties of Bonds to PAA Prepared Surfaces with Different Deoxidizing Processes [29]

Alloy	Deoxidizer	Floating Roller Peel Strength (pli)		Lap Shear Strength (psi)		
		−67°F	75°F	−67°F	75°F	160°F
2024 T-6	Brush 1	40	73	7,110	6,280	3,800
	Brush 2	54	66	7,100	6,690	4,100
	FPL etch	50	77	7,050	6,540	4,050
7075 T-6	Brush 1	53	70	9,000	6,700	4,700
	Brush 2	53	69	8,450	6,200	4,300
	FPL etch	51	71	9,120	6,850	4,740

Table 2.2 Physical Properties of Bonds to PAA Surfaces After Environmental Exposure [29]

Condition A−140°F and 100% RH Condition B−Salt Spray

Alloy	Deoxidizer	Floating Roller Peel Strength (pli)		Lap Shear Strength (psi)	
		Condition A	Condition B	Condition A	Condition B
2024 T-3	Brush 1	80	78	6,125	6,375
	Brush 2	75	72	6,125	6,375
	FPL etch	76	73	5,800	6,150
7075 T-6	Brush 1	70	69	6,700	7,150
	Brush 2	69	71	6,360	6,100
	FPL etch	72	71	7,150	7,000

brushes to abrade the surface of the aluminum prior to anodizing. The results are summarized in Tables 2.1 and 2.2.

Comparison of the data indicates that there is no difference between the surfaces prepared using abrasive deoxidation or the sulfuric acid-sodium dichromate (FPL) deoxidation process prior to the PAA treatment. Rogers [31] had also reported that the FPL deoxidation step was not a critical step in the PAA process but could be replaced by the P2 etch. Adelson and Wegman [32] showed that when the P2 etch was used to replace the FPL etch as the deoxidizer in the PAA process, the durability of adhesive bonds to both 2024 T-3 and 5052 H-34 aluminum was as good. This is shown in Table 2.3.

Table 2.3 Durability Data–PAA Treated Aluminum [32]

Adherend 2 2024 T-3
Adherend 5 5052 H-34
Adhesive A Hysol EA-9628NW with American Cyanamid BR 127 primer
Adhesive B Narmco Metlbond M1117 with American Cyanamid BR 127 primer

Process	Adherend/Adhesive	Time to Failure (hr) Stress				
		70%	60%	50%	40%	30%
PAA	2/A	NR*	149	277	938	713
P2/PAA	2/A	NR	113	347	406	656
PAA	2/B	361	1,107	2,747	4,549	NR
P2/PAA	2/B	537	907	3,003	5,992+	NR
PAA	5/B	2,307+	NR	NR	NR	NR
P2/PAA	5/B	2,169+	NR	NR	NR	NR

*NR signifies test data was not reported at these levels.

2.4 P2 Etch Process

The P2 etch was developed as a modification of the P etch which was first presented in 1976 by Russell [30]. The intent behind the development of the P etchants was to develop a surface treatment for aluminum which would be at least equal to the industrial standard, the FPL etch. The new etchant was to be chromate free and have minimal toxicity. The original P etch developed by Russell contained a significant quantity of nitric acid. When aluminum is treated in the P etch the reaction results in the production of various oxides of nitrogen. These oxides, besides being objectionable, are toxic and require special venting and scrubbers. The P2 etch was then developed by Russell and Garnis [7]. This etchant contains 370 grams of concentrated sulfuric acid, 150 grams of 75% ferric sulfate, and sufficient water to produce one liter of the etchant.

Based upon a preproduction evaluation, Rogers [31] reached the following conclusions. The P2 etch process does not degrade the properties of the aluminum alloys beyond acceptable levels. It produces surfaces that are receptive to adhesives. It provides surfaces to which adhesives bond with strengths which are equal to or better than those produced by the FPL etch. It is an acceptable pretreatment (deoxidizer) prior to either the phosphoric acid anodize (PAA) or the chromic acid anodize (CAA), and when used alone or as part of the PAA process will provide a chromate-free surface treatment for the preparation of aluminum alloys for adhesive bonding. Rogers also stated that the P2 etch process does not require elaborate waste disposal procedures as is the case of solutions containing hexavalent chromium compounds.

Quick [33] evaluated the effect of cure temperature and pressure on the adhesive bond properties of P2 etched aluminum and reported that the chromate-free, low toxicity P2 etch was definitely worthy of consideration as a surface treatment for aluminum. The lap shear and peel strengths of the P2 etched aluminum were comparable to those of the PAA and the optimized FPL etch treated surfaces. In 1984 a team of investigators at the U.S. Army Armament Research and Development Center [34] published a paper on the functioning of the P2 etch in treating aluminum alloys for adhesive bonding. The discussion focused primarily on the 2024 aluminum alloy although other alloys were not excluded. The authors stated that the 2024 aluminum alloy is believed to be attacked by sulfuric acid as shown in the following equations:

$$2AI + 6H^{+} \rightarrow 2AI^{+++} + 3H_2 \quad (2.3)$$

$$Cu + 4H^{+} + SO_4^{=} \rightarrow Cu^{++} + SO_2 + H_2O \quad (2.4)$$

Equation (2.3) is the equation for the standard attack on aluminum by acids and is known to occur. Equation (2.4) shows that hot sulfuric acid acts as an oxidizing agent by attacking the copper. This reaction was verified by the authors by treating copper plate in a P2 etch solution which did not contain ferric sulfate. The solution slowly dissolved the copper as indicated by the characteristic blue color of Cu^{++}. The reaction of ferric ion with the aluminum and copper is shown by Equations (2.5) and (2.6).

$$3Fe^{+++} + Al \rightarrow Al^{++} + 3Fe^{++} \quad (2.5)$$

$$2Fe^{+++} + Cu \rightarrow 2Fe^{++} + Cu^{++} \quad (2.6)$$

Equation (2.5) is based on the fact that ferric salts are corrosive to aluminum [35]. Further, the authors point out that acids carrying oxidizing agents such as ferric salts will attack copper. This indicates that Equation (2.6) is based on firm ground [36]. The authors propose that the ferric ion in the P2 etch retards the attack of the sulfuric acid on the surface of the aluminum as demonstrated by the slow changes in the concentration of the acid. At the same time the surface is pitted by the reactions of the ferric ion as shown in Equations (2.5) and (2.6). Samples treated with sulfuric acid alone showed only a simple dissolution of the aluminum, whereas, when the P2 etch was used, much more pitting of the surface was observed.

2.5 Chromic Acid Anodize Process

Chromic acid anodizing has been used to protect aluminum from oxidation for many years. The anodizing process builds a basic oxide layer which supports a tubular cell structure. This structure is then generally sealed in hot water by conversion of the porous cell structure to the trihydrated aluminum oxide. This sealed oxide layer is thick, structurally weak and generally not satisfactory for adhesive

Table 2.4 MMM-A-132 Lap Shear Strength of Adhesive Bonds to Surfaces with Four Different Pretreatments [31]

Process and Exposure	Lap Shear Strength (psi)	
	Control	After Exposure
FPL etch		
Salt spray	4,343	4,995
Water	4,314	4,221
Humidity	4,550	4,407
Fuel	4,684	5,122
Hydraulic	4,824	5,069
Lubricant	4,768	5,296
P2 etch		
Salt spray	4,608	4,716
Water	4,603	4,707
Humidity	4,778	4,106
Fuel	4,634	4,921
Hydraulic	4,491	5,085
Lubricant	4,945	4,962
CAA		
Salt spray	4,368	4,923
Water	4,204	4,348
Humidity	4,410	4,339
Fuel	4,560	5,129
Hydraulic	4,970	4,873
Lubricant	4,682	4,949
PAA		
Salt spray	4,711	5,177
Water	4,205	5,186
Humidity	4,784	5,087
Fuel	4,514	5,268
Hydraulic	4,871	5,605
Lubricant	5,055	5,167

bonding. By the addition of a small amount of chromic acid to the seal water a sealed anodized surface can be developed to which good bonds can be formed. The acid in the seal water dissolves a part of the cell structure and leaves a thin, tightly adhering, strong layer of aluminum oxide to which adhesives adhere and form good durable bonds.

To compare the four methods of preparing aluminum for adhesive bonding, Rogers [31] subjected both lap shear and peel specimens to the various exposure requirements of MMM-A-132 qualification testing. As shown in Tables 2.4 and 2.5, there were no significant differences in the results obtained with surfaces prepared by using the FPL etch, the P2 etch, the PAA process or the CAA process.

Table 2.5 MMM-A-132 Peel Strength of Adhesive Bonds to Surfaces with Four Different Pretreatments [31]

Process and Exposure	Peel Strength (pli)	
	Control	After Exposure
FPL etch		
Salt spray	75	73
Water	75	72
Humidity	85	75
Fuel	75	77
Hydraulic	77	78
Lubricant	85	78
P2 etch		
Salt spray	85	72
Water	78	76
Humidity	75	73
Fuel	75	75
Hydraulic	75	75
Lubricant	75	71
CAA		
Salt spray	80	76
Water	83	79
Humidity	78	72
Fuel	80	79
Hydraulic	77	79
Lubricant	80	75
PAA		
Salt spray	87	86
Water	84	86
Humidity	85	87
Fuel	80	87
Hydraulic	88	84
Lubricant	82	84

2.6 Preparation of Aluminum Alloys by the Sol–Gel Process

The AC-130 surface prebond surface treatment is a Boeing Company Licensed Product Under Boegel-EPH provided by Advanced Chemistry and Technology, Inc. and registered with the U.S. Patent and Trademark Office.

The process is known as sol–gel process and was developed by the Boeing Company as a surface pretreatment for aluminum, titanium, and steel for the repair of aircraft structures at depot and field as well as OEM levels. The process was evaluated by the U.S. Air Force in conjunction with the U.S. Navy, U.S. Army, and the Boeing Company [37,38].

The process is described as a high-performance surface preparation for adhesive bonding of aluminum alloys, steel, titanium, and composites [39].

The sol–gel process AC-130 promotes adhesion as a result of the chemical interaction at the interfaces between the adherend and the adhesive or primer (Figure 2.4).

For aluminum alloys, the data show AC-130 performs comparatively well to the grit-blast silane process and the PAA (Figure 2.5).

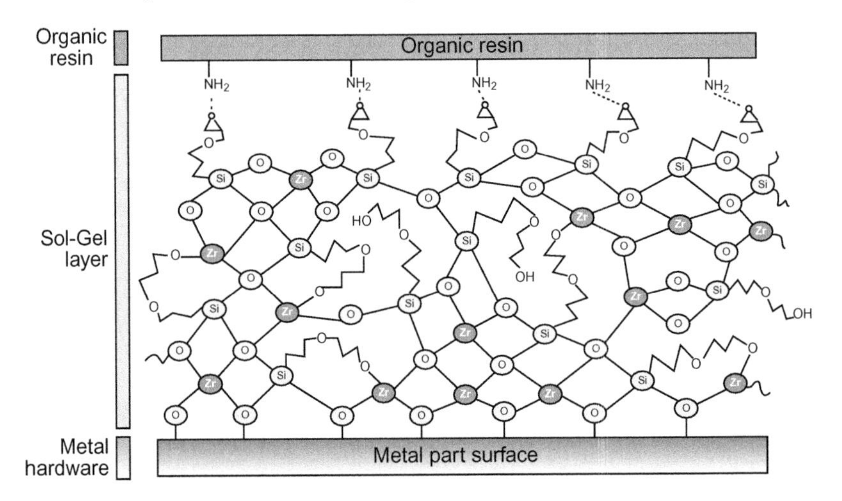

Figure 2.4 Diagram of how sol–gel works.
Source: Courtesy of Advanced Chemistry and Technology, Inc.

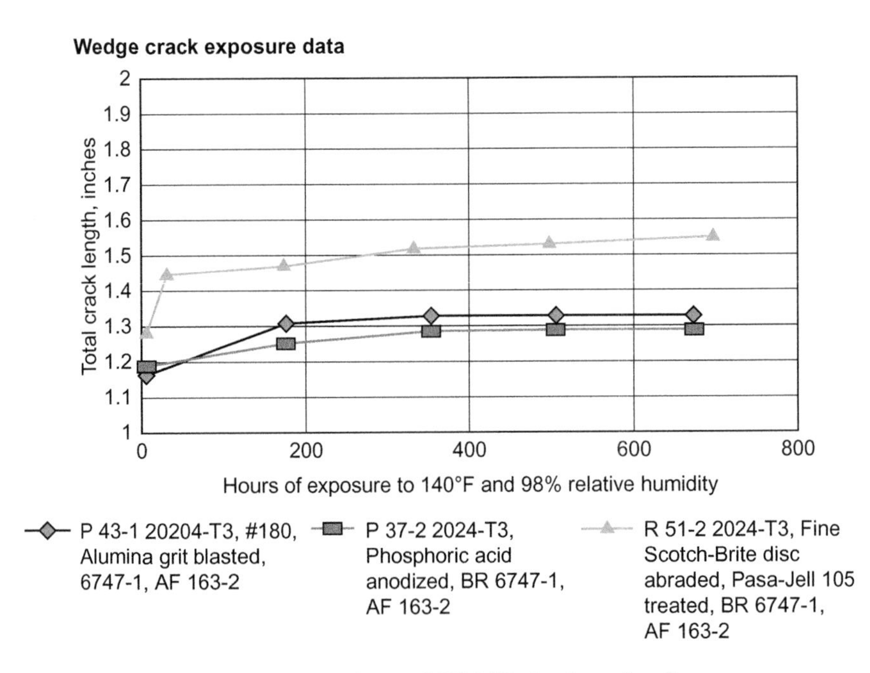

Figure 2.5 Surface treatment comparisons of 2024-T3 aluminum bonding.
Source: Courtesy of Advanced Chemistry and Technology, Inc.

2.7 Preparation of Aluminum Alloys by the Optimized FPL Etch (Sulfo-Chrom Etch)

CAUTION: THE USE OF SODIUM DICHROMATE IS CONSIDERED A HEALTH HAZARD AND IS CONSIDERED A CARCINOGEN. CHECK WITH THE PROPER GOVERNMENTAL AGENCY BEFORE CONSIDERING THE USE OF THIS METHOD.

2.7.1 Apparatus

Treating Tanks

Heated tanks are equipped with automatic temperature controls and have means for agitation to prevent local overheating of the solution. Solutions may be heated by any internal or external means that does not change their composition. Steam should not be introduced into any solution. Compressed air introduced into any solution or equipment must be filtered to remove oil and moisture.

Tank Construction

Tanks are made from, or lined with, materials that have no adverse effect on the solution used or the parts being treated. All tanks should be of a size sufficient to accept the largest part to be processed in a single treatment.

Rinse Tanks

Immersion rinse tanks for potable water are equipped with a means for skimming or overflowing, to remove surface contamination. Deionized water rinse tanks should be fed from the top and discharged from the bottom. The rinse tanks should be large enough to accept the total part in a single rinse.

Transporter/Immersion Racks

Transportor/immersion racks should be equipped with spray or fog heads which will keep the entire part wet during the time of withdrawal from the etch tank until immersion into the rinse tank.

2.7.2 Materials

Rinses

Rinses, other than the final rinse, should be maintained to prevent carryover of materials that would adversely effect the next solution.

Water

Water used for processing solutions and final rinses should meet the requirements listed below.

	Minimum	Maximum
pH	6.0	8.0
Total solids, ppm		200
Total alkalinity as $CaCO_3$, ppm		125
Chloride content, ppm		15

Analyses should be performed as often as necessary to assure that the water meets the requirements. Samples for analysis should be collected at the processing tank.

Degreasing

Oils and greases are to be removed when necessary by vapor degreasing, solvent cleaning, emulsion cleaners or alkaline degreasers. When necessary excess oils and greases are to be removed by emulsion cleaners or alkaline degreasers at room temperature. Vapor degreasing may be accomplished with any of the following solvents, depending upon local regulations, with boiling temperatures to be controlled within the ranges shown:

$$1, 1, 1-\text{trichlorethane—}74^\circ \text{ to } 78^\circ\text{C}(165^\circ \text{ to } 172^\circ\text{F}) \text{ MIL}-\text{T}-81533$$

$$\text{Trichloroethylene—}87^\circ \text{ to } 90^\circ\text{C}(188^\circ \text{ to } 193^\circ\text{F}) \text{ O}-\text{T}-620$$

$$\text{Perchloroethylene—}120^\circ \text{ to } 124^\circ\text{C}(248^\circ \text{ to } 255^\circ\text{F}) \text{ O}-\text{T}-236$$

Emulsion cleaners and alkaline degreaser are to be used in accordance with the manufacturers' recommendations.

Alkaline Cleaning Solution

Non-etch alkaline cleaning solutions are to be prepared in accordance with the manufacturers' recommendations. When the aluminum being cleaned is immersed in the alkaline cleaner for the time and temperature used for processing, there should be no evidence of gas evolution, etching or metal removal. The alkaline cleaner must not contain silicates.

Etch Solution

The chemical analysis of the etch solution should be maintained at approximately 30 pbw water, 10 pbw of sulfuric acid (specific gravity 1.83) and 1 to 4 pbw of sodium dichromate. Prior to use, a minimum of 0.25 oz/gal of 2024 aluminum should be dissolved in the solution.

2.7.3 Processing

Degreasing

Remove any markings by solvent wiping using safety solvent. Remove any visible oil and/or grease from the aluminum by vapor degreasing or solvent cleaning using the materials specified in Section 2.7.3. Remove the parts being degreased from the degreaser when condensation ceases. Degreasing may be repeated as needed.

Alkaline Cleaning

Unless otherwise recommended by the manufacturer, immerse the aluminum for a minimum of five minutes, in the alkaline cleaning solution, which should be held at 50° to 80°C (122° to 176°F). Follow the alkaline cleaning by a thorough rinsing in water held at a temperature of 32° to 70°C (75° to 158°F). Alkaline cleaning may be repeated as necessary. Keep the parts wet between the alkaline treatment and immersion in the rinse tank.

Etching

Immerse the aluminum parts in the etch solution for nine to fifteen minutes at a temperature from 65° to 70°C (149° to 158°F). Keep the parts wet between the etch tank and the rinse tank. Follow the etching by a thorough rinsing with room temperature water. (See ASTM E171 for the definition of room temperature.) Parts that are rerun, because of lack of water break or stains, or unprimed parts that have exceeded the allowable storage time, may be reworked no more than twice. A total immersion time of 34 minutes should not be exceeded.

Final Rinse

Rinse the aluminum for one to two minutes in water at a temperature not to exceed 50°C (122°F).

Drying

Air dry the aluminum parts for not more than one hour at a temperature not to exceed 65°C (149°F).

2.7.4 Restrictions

Parts should be completely immersed into all solutions. Etching and rinsing stains on the bonding surfaces are not acceptable. Cleaned parts which are not to be primed should be placed in a controlled area within thirty minutes of the completion of the cleaning. The controlled area should have air that is maintained at a maximum relative humidity of 50%. Cleaned parts should be primed or bonded within 24 hours of cleaning. Parts that have been processed should be handled with care using clean white cotton gloves.

2.8 Preparation of Aluminum Alloys by the Phosphoric Acid Anodize Process (PAA)

2.8.1 Apparatus

Treating Tanks

Heated tanks are to be equipped with automatic temperature controls and should have means for agitation to prevent local overheating of the solution. Solutions may be heated by any internal or external means that does not change their composition. Steam should not be introduced into any solution. Compressed air introduced into any solution or equipment must be filtered to remove oil and moisture.

Tank Construction

Alkaline Cleaning Tank—The suggested tank material is mild steel. *Deoxidizing Tank*—The deoxidizing tank should be made from or lined with materials that have no adverse effect on the solution used or the parts being treated. All tanks should be big enough to accept the largest part to be processed in a single treatment. *Rinse Tanks*—Immersion rinse tanks for tap water should be equipped with a means for skimming or overflowing, to remove surface contamination. For deionized water the rinse tanks should be fed from the top and discharged from the bottom. The rinse tanks should be large enough to accept the total part in a single rinse. *Anodizing Tanks*—The anodizing tank should be constructed of commercial lead with 6 to 8% Sb or stainless steel (347 or 316 or equal). It should be equipped with an oilless filtered air agitation system. The anodizing racks and frames should be constructed of materials that will not cause corrosive reactions with the suspension wires, clips or the parts during the entire process. Parts are to be attached to the racks or frames with 1100 S aluminum alloy wires or hooks, or with titanium alloy spring clips.

Electrical System

The electrical system should be adequate to provide 10 volts dc and to maintain any set voltage within 0.1 volt dc within a two to three minute time span.

2.8.2 Appearance

The phosphoric acid anodize process produces a coating which is continuous, smooth, uniform in appearance and when examined visually, should be free from discontinuities, such as scratches, breaks, burned areas and areas which are not anodized. There should be no stains, streaks, discoloration or residue on the surface of the anodized parts after rinsing. During rinsing and draining there should be no evidence of water breaks.

Color

The anodized coating on the parts may appear light gray with some irridesence.

Racking Marks

Racking marks are acceptable provided there is no evidence of burning. Burns can be identified by pitted or melted surfaces at the clamp marks. Burns of any degree that will be in the final part area are cause for rejection. Non-current-carrying supports that mark the surface should be held to a minimum. If possible the current clamps should be placed in a nonbonding area of the part.

2.8.3 Materials

Vapor Degreasing/Solvent Cleaning

When necessary excess oils and greases are to be removed by vapor degreasing, solvent cleaning, emulsion cleaners, or alkaline degreasers.

Alkaline Cleaning

The alkaline cleaning solution should contain 0.37 to 0.43 lb/gal of TURCO 4215 Special in deionized water plus 0.0075 gal of TURCO 4215 Additive for each pound of TURCO 4215 Special.

Deoxidizing Solution

The deoxidizing solution is the FPL etch solution. See the FPL process for the formulation. When the FPL process is not approved because of the dichromate restriction the P2 etch process may be used as a deoxidizer.

Anodizing Solution

The anodizing solution contains 9 fluid ounces of 85% phosphoric acid per gallon of deionized water.

2.8.4 Processing

Degreasing

Remove any marks by solvent wiping using safety solvent. Remove any visible oil and grease from the surface of the aluminum as specified in Section 2.8.4. Remove the parts being degreased from the degreaser as soon as condensation ceases. Degreasing may be repeated as needed.

Alkaline Cleaning

The parts are to be cleaned at 66° ± 2°C (150° ± 5°F) in the solution specified in Section Alkaline Cleaning.

Deoxidation

The parts are deoxidized in the FPL etch solution by immersion in the solution for two to ten minutes at 60° to 71°C (140° to 160°F).

Rinse

The parts are rinsed in tap water at room temperature.

Anodizing

The parts are made the anode in the solution specified in Anodizing Solution. The cathode is to be made of 347 or 316 stainless steel. Apply a dc voltage stepwise to 10 ± 1 volts in two to five minutes. Maintain at 10 ± 1 volts for 20 to 25 minutes. Turn off the current.

Rinse

Within two minutes after the current has been turned off, immerse the parts into agitated and overflowing clean tap water for ten to fifteen minutes.

Stripping

If necessary, the anodic coating may be stripped and reworked by returning the part to the deoxidizing solution for a period of two to three minutes and then repeating the subsequent steps.

2.8.5 Restrictions

Parts should be primed with a corrosion-inhibiting primer within 120 minutes after drying.

2.8.6 Anodic Coating Inspection

The surface of the part is illuminated by using a mercury vapor or fluorescent lamp. The surface is then observed at a low angle (0 to 10 degrees) through a photographic polarizing filter held close to the viewer's eye. The anodized surface should display interference colors which change to a complementary color when the filter is rotated 90 degrees (e.g., from purple to yellow-green). Different aluminum alloys will show different colors. The colors most frequently seen will be purple, yellow, blue and green hues. Abrupt differences in color level area, except at electrical contact points, from background colors are not acceptable. Causes of such differences may be fingerprints, abrasion or other contamination.

2.9 Preparation of Aluminum Alloys by the P2 Etch (Sulfo-Ferric Etch)[1]

2.9.1 Apparatus

Treating Tanks

Heated tanks are to be equipped with automatic temperature controls and should have means for agitation to prevent local overheating of the solution. Solutions may be heated by any internal or external means that does not change their composition. Steam is not introduced into any solution. Compressed air introduced into any solution or equipment must be filtered to remove oil and moisture.

Tank Construction

Tanks shall be made from or lined with materials that have no adverse effects on the solution used or the parts being treated. All tanks must be big enough to accept the largest part to be processed in a single treatment.

Rinse Tanks

Immersion rinse tanks shall be equipped with a means for skimming or overflowing or both to remove surface contamination. The tanks shall be equipped with a means for flushing hollow sections.

Transporter/Immersion Racks

Transporter/immersion racks should be equipped with spray or fog heads which will keep the entire part wet during withdrawal from the etch tank until immersion into the rinse tank.

Rinses

Rinses other than the final rinse must be maintained in such a manner to prevent carryover of material that would adversely effect the next solution.

[1] See ASTM E864 for possible update for the P2 process.

2.9.2 Materials

Water

Water used for makeup of processing solutions and final rinses should meet the following requirements:

	Minimum	Maximum
pH	6.0	8.0
Total solids, ppm		200
Total alkalinity, as $CaCO_3$, ppm		125
Chloride content, ppm		15

Analyses must be performed as often as necessary to assure that the water meets these requirements. Samples for analysis should be collected at the processing tank.

Alkaline Cleaning Solution

Non-etch alkaline cleaning solutions should be prepared and used in accordance with the manufacturer's recommendations or as indicated in Section 2.9.3. When the aluminum being cleaned is immersed in the alkaline cleaner for the time and at the temperature used for processing, there should be no evidence of gas evolution, etching or metal removal. The alkaline cleaner should not contain silicates.

Etch Solution

The chemical analysis of the etch solution shall be maintained at approximately 27 to 36% by weight of sulfuric acid (specific gravity 1.84 and 2.9 to 4.7 oz/gal (22 to 35 g/l) of ferric iron or 18 to 22 oz/gal (135 to 165 g/l) of ferric sulfate). This is the equivalent to 2 gal (7.56 l) of concentrated sulfuric acid and 12.5 lbs (5.69 kg) of ferric sulfate in every 10 gal (37.8 l) of solution. Two gal (7.56 l) of a 50% ferric sulfate solution can be used in place of the 12.5 lbs (5.69 kg) of powdered ferric sulfate. Note: Only virgin ferric sulfate solution shall be used in this process. Impurities in reclaimed ferric sulfate will cause unwanted reactions when the aluminum is treated.

2.9.3 Processing

Alkaline Cleaning

Unless otherwise recommended by the manufacture, immerse the aluminum, for a minimum of five minutes, in the alkaline cleaning solution, which is to be held at 50° to 80°C (122° to 176°F). Follow the alkaline cleaning by a thorough rinsing in water held at 23° to 70°C (75° to 158°F). Alkaline cleaning may be repeated as necessary. Keep the parts wet between the alkaline treatment and immersion in the rinse tank.

Etching

Immerse the aluminum parts in the etch solution for 10 to 12 minutes at a temperature from 60° to 65°C (140° to 149°F). Keep the parts wet between the etch tank and the rinse tank. Follow the etching by thorough rinsing with room temperature water.

Parts that are to be rerun, because of lack of water break, stains, or unprimed parts that have exceeded the permitted storage time, may be reworked no more than twice. Do not exceed a total immersion time of 34 minutes.

Final Rinse

Rinse the aluminum for one to two minutes in water from room temperature to 50°C (122°F).

Drying

Air dry the aluminum for not more than one hour at a temperature not to exceed 65°C (149°F).

2.9.4 Restrictions

Parts should be completely immersed in all solutions. Etching and rinsing stains on the bonding surface are not acceptable. Cleaned parts which are not to be primed are to be placed into a controlled area within thirty minutes of the completion of the cleaning. The controlled area should have air that is maintained at a relative humidity of 50% maximum. Cleaned parts should be primed or bonded within 24 hours of cleaning. Parts that have been processed should be handled with care using clean white cotton gloves.

2.10 Preparation of Aluminum Alloys by the Chromic Acid Anodize Method (CAA)

CAUTION: THE USE OF CHROMIC ACID IS CONSIDERED A HEALTH HAZARD AND IS CONSIDERED A CARCINOGEN. CHECK WITH THE PROPER GOVERNMENTAL AGENCY BEFORE CONSIDERING THE USE OF THIS METHOD.

2.10.1 Apparatus

Treating Tanks

Heated tanks should be equipped with automatic temperature controls and have means for agitation to prevent local overheating of the solution. Solutions may be heated by any internal or external means that will not change their composition.

Steam must not be introduced into any solution. Compressed air introduced into any solution or equipment must be filtered to remove oil and moisture.

Tank Construction

Alkaline Cleaning Tank—The suggested tank material is mild steel. *Deoxidizing Tank*—The deoxidizing tank should be made from or lined with material that has no adverse effect on the solution used or the parts being treated. Tanks must be big enough to accept the largest part to be processed in a single treatment. *Rinse Tanks*—Immersion rinse tanks for potable tap water should be equipped with a means for skimming or overflowing to remove surface contamination. Deionized water rinse tanks should be fed from the top and discharged from the bottom. The rinse tanks must be large enough to accept the total part in a single rinse. *Anodizing Tank*—The anodizing tank is to be of commercial lead with 5 to 8% Sb and be equipped with an oil-free filtered air agitation system. The anodizing racks and frames are to be constructed of material that will not cause corrosive reactions with clips, suspension wires or the parts during the entire process. Parts are to be attached to the racks or frames with 1100 series aluminum wire or hooks. Titanium alloy spring clips are also acceptable.

Electrical System

The electrical system is to be adequate to provide 40 volts dc and to maintain any set voltage within ± 1 volt within a two to three minute time span.

2.10.2 Appearance

Color

The anodized coating on 2024 aluminum alloy is normally dark gray. In the case of 2014 and 5052 aluminum alloys, the coating often appears transparent. Other alloys may have coatings of various shades and the user will have to determine these by experimentation. Extrusions and forgings will show varying amounts of grain structure, depending on the heat treatment and condition of the alloy.

Racking Marks

Racking marks are acceptable provided there is no evidence of burning. Burns can be identified by a pitting or melted surface at the clamp mark. Burns, of any degree, that will be in the final part are cause for rejection. Noncurrent carrying supports that mask the surface should be held to a minimum and should be placed away from the edge of the part in so far as possible. Heavy chromate stains that occur at racking points may be removed by blotting with wet cheesecloth.

Wiping

Anodized surfaces are not to be wiped with dry cheesecloth, gloves or other cloths. Care should be exercised to prevent any wiping or rubbing during inspection, handling and lay-up.

2.10.3 Materials

Water

Water used for makeup of processing solutions and final rinses shall meet the following requirements:

	Minimum	**Maximum**
pH	6.0	8.0
Total solids, ppm		200
Total alkalinity, as $CaCO_3$, ppm		125
Chloride content, ppm		15

Analyses are to be performed as often as necessary to assure that the water meets these requirements. Samples for analysis are to be collected at the processing tank.

Degreasing

When necessary excess oils and greases are to be removed by use of emulsion cleaners or alkaline degreasers. Emulsion cleaners and alkaline cleaners shall be used in accordance with the manufacturers' recommendations.

Alkaline Cleaning Solutions

Non-etch alkaline cleaning solutions are to be prepared in accordance with the manufacturers' recommendations. When the aluminum being cleaned is immersed in the alkaline cleaner for the time and at the temperature used for processing, there should be no evidence of gas evolution, etching or metal removal. The alkaline solution should not contain silicates.

Deoxidation Solution

The deoxidation solution is to be made up with approximately 30 pbw water, 10 pbw sulfuric acid (specific gravity 1.84), and 1 pbw sodium dichromate. Note: the P2 etch may be used as the deoxidation solution.

Anodizing Solution

The anodizing solution should be made up with flake chromic acid and water. It should be maintained at 6 to 10% by weight total chromic acid, 3% by weight minimum free chromic acid, pH of 0.8 maximum, 0.03% by weight maximum chloride content, and 0.05% by weight maximum sulfate content.

Seal Solution

The seal solution is to be made up with 75 to 120 ppm chromic acid in water; the pH being held between 2.5 and 3.8 when checked at 25°C (77°F).

2.10.4 Processing

Degreasing

Remove any visible oil and grease using the materials in the section Degreasing, above.

Alkaline Cleaning

The parts are to be cleaned for five to ten minutes in accordance with the manufacturer's recommendations (see Section 2.9.3). Alkaline cleaning is to be followed by thorough rinsing in water at room temperature up to 71°C (160°F). Cleaning may be repeated but no single immersion should exceed ten minutes.

Deoxidizing

Deoxidizing should be accomplished by immersion in the solution described in Section Deoxidation Solution at 60° to 71°C (140° to 160°F) for a period of five to ten minutes. The deoxidation is to be followed by thorough rinsing at room temperature.

Anodizing

The parts are made the anode of the electrical system. Contact points, bus bar connections and all other current-carrying connections that are made during loading must be cleaned prior to making the connections to assure positive contact. The anodize solution is gently agitated and is operated at 33° to 37°C (92° to 98°F) during the anodizing process. The voltage is applied stepwise as follows. Apply an initial voltage of 5 to 10 volts dc. Observe a 2 to 2.5 minute time delay between the initial voltage and the addition of a second 5 to 10 volt increase. Increase the voltage by 5 to 10 volts at approximately one minute intervals until a voltage of 40 ± 2 volts is obtained. Not less than five minutes and not more than ten minutes are required to reach the 40 ± 2 volts. The parts are to be processed at 40 ± 2 volts for 30 to 35 minutes. The minimum current density is 1 amp per square foot at 40 ± 2 volts. The anodized part should not remain in the anodize solution for more than five minutes after the current has been turned off.

Rinsing

After anodizing, the parts are rinsed in room temperature water.

Sealing

The anodized parts are sealed in the solution described in the section Seal Solution at 82° to 85°C (180° to 185°F) for seven to nine minutes. The parts are not rinsed following the sealing operation.

Stripping

When stripping is required, it is done by immersion in the deoxidizing solution (see the section Deoxidizing).

2.10.5 Restrictions

Anodizing is not to be used on aluminum alloys that have a nominal copper content in excess of 5% or that have a nominal total alloying content in excess of 7.5%. Parts that are found to be unsatisfactorily anodized, parts that are mechanically reworked and parts that have exceeded the storage time limits should be completely stripped and reanodized. However, parts should not be stripped and reanodized more than twice. Anodized parts should be bonded or primed within thirty days after treatment.

2.10.6 Specific Controls

Anodize Seal

The chromic acid concentration and the pH of the seal solution should be checked at least daily when in use. The temperature of the solution should be checked prior to each load.

Anodic Coating Weight

The anodic coating weight should be checked at least once each month. Specimens should be processed with a production load and tested as follows:

1. Mark the edges of the specimen for identification.
2. Weigh the specimen on an analytical balance and record the weight to the nearest 0.1 mg.
3. Strip the anodic coating in a solution containing 27 ml phosphoric acid and 20 g chromic acid in 1,000 ml distilled water.
4. Immerse the specimen for five minutes at 82° to 99°C (180° to 210°F).
5. Rinse in distilled water, dry and weigh.
6. Repeat until a constant weight is obtained.
7. Calculate the weight loss in mg/sq ft.
8. The anodic coating weight should not be less than 200 nor more than 600 mg/sq ft.

Corrosion Resistance Test

Corrosion resistance should be checked at least once each month. Specimens should have the longitudinal axis transverse to the direction of rolling and must be cut from adjacent areas of the same sheet. These specimens are to be processed with a typical production load and should be tested as follows. Two specimens are placed in a standard 5 or 20% salt spray (fog) test chamber while two other specimens are retained as controls. The test specimens are exposed for 240 hours. The specimen surface is inclined at a 15 to 30 degree angle from the vertical. If, after 240 hours' exposure, there are more than ten isolated pits on each test specimen, they are subjected to mechanical properties testing. There must not be more than a 5% reduction in the tensile strength and not more than a 10% reduction in elongation as compared to the mechanical properties of the control specimens.

2.11 Preparation of Aluminum Alloys by the AC-130 Sol–Gel Process

2.11.1 Materials

1. AC-130 a four-part system or AC-130-2 a two-part system. AC-Tech 77341 Anaconda Ave, Garden Grove, CA92841
2. Wipers, cheesecloth, gauze, or clean cotton cloth
3. Sanding paper/discs 180-grit, or finer aluminum oxide
4. 3 M Scotch-Brite 2 in (5 cm) or 3 in (7.6 cm) medium grit "Roloc" discs
5. Solvents
6. Bonding primer
7. Proper protective equipment, such as protective gloves, respirators, and eye protection, must be worn during these operations.

2.11.2 Facilities Controls

1. Air used in this process shall be treated and filtered so that it is free of moisture, oil, and solid particles.
2. Application of the surface preparation material and primer shall be conducted in an area provided with ventilation.
3. Grinders used shall have a rear exhaust with an attachment to deliver the exhaust away from the part surface.
4. Sanding tools shall have a random orbital movement.

2.11.3 Manufacturing Controls

1. Parts to be processed shall be protected from oil, grease, and fingerprints.
2. Mask dissimilar metals and neighboring regions where appropriate.
3. Apply bond primer within 24 hours of AC-130 application. Cool parts to room temperature.
4. Apply adhesive within 24 hours of AC-130 application.
5. Contain grit and dust residues generated during mechanical deoxidization processes.

2.11.4 Storage

1. Sol–gel kits are considered to be time- and temperature-sensitive and shall be stored in accordance with the supplier's recommendations.

2.11.5 Processing

1. **Precleaning**: Remove all foreign materials from the surface as needed.
2. Prepare the AC-130 or 130-2 in accordance with the manufacturer's instructions provided in each kit. Scale up for the size of the part as necessary, e.g., 1 liter of solution per 0.9 m^2 (10 ft^2) to be coated. Do not treat the surface with the AC-130 or 130-2 before 30 min induction time is complete. (Induction time is defined as the time period after all the AC-130 components have been mixed, but before the mixed solution is active.) Do not treat surfaces after the 10-hour maximum pot-lifetime has expired.
3. Deoxidation of the surface may be accomplished by grit blasting, sanding, mechanical Scotch-Brite, or manual Scotch-Brite.
   a. **Grit Blasting**: Using alumina grit, grit blast an area slightly larger than the bond area. Use 2–5 atm (30–80 psig) oil-free compressed air or nitrogen. Slightly overlap the blast area with each pass across the surface until a uniform matte appearance has been achieved.
   b. **Sanding**: Using a sander or high-speed grinder connected to oil-free nitrogen or compressed air line, thoroughly abrade the surface with abrasive paper for 1–2 min over a 15 × 15 cm (6 × 6 in) section covering the entire surface uniformly, following the manufacturer's instructions [40].
   c. **Mechanical Scotch-Brite Abrading**: Replace the abrasive paper as described in Section 2.11.1.3 with a Scotch-Brite abrasive disc and abrade surface as described in Section 2.11.1.4.
   d. **Manual Scotch-Brite**: Thoroughly abrade the surface with a very fine Scotch-Brite pad for a minimum of 1–2 min as described in Section 2.11.1.3. Remove loose grit residue with a clean, dry, natural bristle brush or with clean oil-free compressed air or nitrogen.
4. **Application of AC-130**: Apply AC-130 or 130-2 solution as soon as possible after completion of the deoxidization process. The time between completion of deoxidization and application of AC-130 shall not exceed 30 min. The application of the AC-130 or 130-2 may be accomplished by any of the following methods: spray application, manual application using a clean natural bristle brush or swabbing with a clean wiper, cheesecloth or gauze, or by immersion. Apply solution generously, keeping the surface continuously wet with the solution for a minimum period of one minute. Surface must not be allowed to dry and should be covered with fresh solution at least one time during the solution application period. Allow the coated surface to drain for 5–10 min. If there is any surplus AC-130 solution collected in crevices, pockets, or other contained areas, use filter compressed air to lightly blow off excess solution while leaving a wet film behind. Do not splatter this excess solution onto adjoining surfaces.
5. **Drying**: Allow the coated part to dry under ambient conditions for a minimum of 60 min. Minimize contact with the part during this time, as the coating may be easily damaged until fully cured. Exact drying time will depend on the configuration of the part and room conditions (temperature and humidity).
6. **Bonding**: Apply primer and/or adhesive within 24 hours of AC-130 or 130-2 application. Keep the surface clean during the entire operation.

2.11.6 Acceptable Results

An acceptable AC-130 or 130-2 coating is smooth and continuous without evidence of surface contamination or defects. Dark areas caused by draining off of the sol–gel solutions are acceptable.

References

[1] H.W. Eickner, N.E. Schowalter, A Study of Methods for Preparing Clad 24 S-T3 Aluminum Alloy Sheet Surfaces for Adhesive Bonding, Forest products laboratory report no. 1813 (1950).
[2] A.W. Bethune, Durability of bonded aluminum structure, SAMPE J 11 (3) (1975) 4–10.
[3] R.H. Olsen, Implementation of phosphoric acid anodizing at air logistic centers, SAMPE NSTC 11 (1979) 770–779.
[4] J.C. McMillan, J.T. Quinlivan, R.A. Davis, Phosphoric acid anodizing of aluminum for structural bonding, SAMPE Q 7 (3) (1976) 13–18.
[5] H.S. Schwartz, Effect of adherend surface treatment on stress durability of adhesive bonded aluminum alloys, SAMPE J 13 (2) (1977) 2–13.
[6] ASTM D3933, Standard practice for preparation of aluminum surfaces for structural bonding (phosphoric acid anodizing).
[7] W.J. Russell, E.A. Garnis, A chromate-free low toxicity method of preparing aluminum surfaces for adhesive bonding, SAMPE J 17 (3) (1981) 19–23.
[8] R.F. Wegman, M.C. Ross, S.A. Slota, E.S. Duda, Evaluation of the Adhesive Bonding Processes Used in Helicopter Manufacture Part 1. Durability of Adhesive Bonds Obtained as a Result of Processes Used in the UH-1 Helicopter, Picatinny arsenal technical report 4186, and Bell Helicopter Process Specification FW 4352 Rev E Method 1A (1971) 20.
[9] J.M. Chen, T.S. Sun, J.D. Venables, R. Hopping, Effect of fluorine contamination on the microstructure and bondability of aluminum surfaces, SAMPE NSS&E 22 (1977) 25–46.
[10] J.D. Venables, D.K. McNamara, J.M. Chen, B.M. Ditchek, T.I. Morgenthales, T.S. Sun, R.L. Hopping, Effect of moisture on adhesively bonded aluminum structures, SAMPE NSTC 12 (1980) 909–923.
[11] D.K. McNamara, J.D. Venables, R.L. Hopping, L. Cottrell, FPL etch solution lifetime, SAMPE NSTC 13 (1981) 666–675.
[12] S.P. Kodali, R.C. Curley, L. Cottrell, B.M. Ditchek, D.K. McNamara, J.D. Venables, Effect of rinse water pH on adhesive durability, SAMPE NSTC 13 (1981) 676–684.
[13] L.F. Cottrell, D.L. Trawinski, S.P. Kodali, R.C. Curley, D.K. McNamara, J.D. Venables, Seeding of FPL solution, SAMPE NSS&E 27 (1982) 44–52.
[14] D.K. McNamara, L.J. Matienzo, J.D. Venables, J. Hattater, S.P. Kodali, The effect of rinse water pH on the bondability of FPL-pretreated aluminum surfaces, SAMPE NSS&E 28 (1983) 1143–1154.
[15] J.A. Marceau, Y. Moji, J.C. McMillan, A wedge test for evaluating adhesive bonded surface durability, SAMPE NSS&E 21 (1976) 332–355.
[16] E. Thrall, R. Shannon, PABST surface treatment and adhesive selection, SAMPE NSS&E 21 (1976) 1004–1014.
[17] J.A. Marceau, A SEM analysis of adhesive primer oriented bond failures on anodized aluminum, SAMPE Q 9 (4) (1978) 1–7.
[18] J.D. Minford, Etching and anodizing pretreatments and aluminum joint durability, SAMPE Q 9 (4) (1978) 18–27.
[19] D.K. McNamara, J.D. Venables, T.S. Sun, J.M. Chen, R.L. Hopping, Prebond handling of aluminum surfaces for adhesive bonding, SAMPE NSTC 11 (1979) 740–751.

[20] J.S. Ahern, T.S. Sun, C. Froede, J.D. Venables, R. Hopping, Development of oxide films on Al with the phosphoric acid anodization process, SAMPE Q 12 (1) (1980) 39–45.
[21] D.L. Trawinski, S.P. Kodali, R.C. Curley, D.K. McNamara, J.D. Venables, The effect of fluoride contamination on the durability of PAA surfaces, SAMPE NSTC 14 (1982) 293–301.
[22] E.W. Thrall Jr., Prospects for bonding primary aircraft structures in the 80's, SAMPE NSS&E 25 (1980) 716–727.
[23] E.A. Ledbury, A.B. Miller, P.D. Peters, E.E. Peterson, B.W. Smith, Microstructural characterization of adhesively bonded joints, SAMPE NSTC 12 (1980) 935–949.
[24] D.B. Arnold, E.E. Peterson, Interfacial characteristics of 350°F adhesive bonding, SAMPE NSTC 13 (1981) 162–176.
[25] E.E. Peterson, D.B. Arnold, M.C. Locke, Compatibility of 350°F curing honeycomb adhesives with phosphoric acid anodizing, SAMPE NSTC 13 (1981) 177–188.
[26] K.K. Knock, M.C. Locke, Correlation of surface characterization of phosphoric acid anodize oxide with physical properties of bonded specimens, SAMPE NSTC 13 (1981) 445–458.
[27] O.-D. Hennemann, W. Brockman, Weak boundary zones in metal bonds, SAMPE NSTC 14 (1982) 302–309.
[28] G.D. Davis, J.S. Ahern, J.D. Venables, Hydration of aluminum adherends as studied by surface behavior diagrams, SAMPE NSTC 15 (1983) 202–211.
[29] A.V. Pocius, J.J. Claus, Replacement of sulfuric-chromic acid (FPL) etch by 3-dimensional abrasive surface conditioning for pretreatment of aerospace alloys before H_3PO_4 anodization, SAMPE NSTC 13 (1981) 629–639.
[30] W.J. Russell, A chromate-free process for preparing aluminum for adhesive bonding, in: M.J. Bodnar (ed.), Durability of Adhesive Bonded Structures, J. Appl. Polym. Sci. Applied Polymer Symposium 32, Interscience Publication, John Wiley & Sons, NY (1977) pp. 105–117.
[31] N.L Rogers, Pre-production evaluation of a nonchromated etchant for preparing aluminum alloys for adhesive bonding, SAMPE NSTC 13 (1981) 640–650.
[32] K.M. Adelson, R.F. Wegman, Evaluation of the P2 and PAA treatments, SAMPE NSTC 14 (1982) 311–322.
[33] S. Quick, Effect of cure temperature and pressure on adhesive bond properties of P2 etched aluminum, SAMPE NSS&E 28 (1983) 1116–1126.
[34] R.F. Wegman, D.W. Levi, K.M. Adelson, M.J. Bodnar, The function of the P2 etch in treating aluminum alloys for adhesive bonding, SAMPE NSS&E 29 (1984) 273–281.
[35] T. Lyman (Ed.), Metals Handbook, eighth ed., Properties and selection of metals, American Society for Metals, 1 (1961) 932.
[36] Ibid, 988.
[37] Sol-gel technology for low-VOC, nonchromated adhesive bonding applications SERDP Project PP-1113 task 1 James J. Mazza – AFRL/MLSA, Georgette B. Gaskin – U.S. Navy, William S. De Piero – U.S. Army and Kay Y. Blohowiak – The boeing company, April (2004).
[38] Improvements in surface preparation methods for adhesive bonding Kay Blohowiak SERDP/ESTCP Workshop, Tempe AZ, USA (2008).
[39] A-C tech literature AS-130 metal prebond surface treatment.
[40] A-C tech application guide AC-130 and AC-130-2 metal alloy surface preparation for bonding.

3 Titanium and Titanium Alloys

3.1 Introduction

Titanium and its alloys are useful because of their high strength, light weight, and ability to withstand elevated temperatures. The combination of these properties and adhesive bonding in properly designed structures have allowed design engineers to design and build various advanced, high performance structures. During the latter part of the 1950s, various methods were used to treat titanium for adhesive bonding. One of these early treatments was an anodic process which produced a surface to which adhesives would adhere well. The joints, however, would fail due to failure of the anodic coating to adhere to the metal surface. Alkaline cleaning techniques were then introduced. The alkaline cleaned surfaces resulted in good initial bonds, but the joints failed in relatively short periods of time. During the early sixties both alkaline cleaning and a nitric-hydrofluoric acid pickling process were used. In 1966 and in 1967, Allen and Allen [1,2] reported that two new techniques had been discovered to produce fluoride conversion coatings on titanium metal and on 6 Al-4 V-titanium. In 1966 a phosphate-fluoride process for treating titanium was in use. This process appeared to be superior to those previously used.

In 1968 a team of investigators at Picatinny Arsenal embarked on an investigation to determine how much better the phosphate-fluoride process was and to attempt to determine the life expectancy of adhesive bonds produced by this process [3,4]. The investigators reported [4] that "both the alkaline cleaning and the phosphate-fluoride prebond treatments were found to produce surfaces of good wettability, equal thickness and similar composition." The major difference between the surfaces left by the two processes appeared to be in the structure of the titanium dioxide produced. The authors reported that on the alkaline cleaned surfaces the titanium dioxide was in the rutile form. This is the more stable form of titanium dioxide and is the same form found on the as-received metal. The phosphate-fluoride treated surfaces appeared to contain the anatase structure of titanium dioxide. Aged unbonded phosphate-fluoride treated titanium surfaces and aged bonded joints of phosphate-fluoride treated titanium which had failed showed some rutile structure.

It was concluded that the anatase structure on the titanium reverted to the thermodynamically more stable rutile and that this conversion involved an 8% change in volume. This conversion and the accompanying volume change would be expected to have an effect on long term bond stability. In 1973, Wegman and Bodnar [5] introduced a stabilized phosphate-fluoride process which was intended

Surface Preparation Techniques for Adhesive Bonding.

to retard the conversion of anatase to rutile. Other studies have been published covering this process [6–9].

In 1976, Marceau, Moji and McMillan [10] reported that Boeing had developed an anodic process which produced a porous oxide structure that was somewhat similar to that which is produced on aluminum, and which was consistently superior to the surface formed by the "state-of-the-art" phosphate-fluoride process. This anodic process is described in a U.S. patent [11]. Locke, Harriman and Arnold [12] optimized the Boeing process in 1980 and compared the process to other processes used in the industry, including such processes as the Pasa-Jell 107, VAST, Turco 5578 and the Picatinny Arsenal modified phosphate-fluoride processes. Other investigations [13–17] were being conducted on various processes for treating titanium and its alloys. With interest in higher temperature applications there was an increasing desire to attempt to classify and determine the best surface preparation processes.

One of the most comprehensive investigations into the methods of treating titanium was initiated by the U.S. Naval Air Systems Command in 1978. A team of investigators from the U.S. Naval Air Development Center, the U.S. Army Armament Research and Development Center, and the Martin Marietta Laboratories, working with the cooperation of eight aerospace companies and three adhesive manufacturers, evaluated the leading methods for treating titanium alloys prior to adhesive bonding. The data obtained during this evaluation have been presented in various papers and reports [15,18–26]. During this investigation different surface prebond treatments were evaluated. These included the phosphate-fluoride process, modified phosphate-fluoride processes, the Dapcotreat process, the dry hone/Pasa-Jell 107 process, the Turco 5578 process, chromic acid anodic processes and the alkaline-peroxide process. A study of the morphology and decomposition of the various surfaces [15] by use of scanning transmission electron microscopy along with ESCA/Auger techniques revealed that very different surface morphologies are obtained with the different treatments. The investigators were able to classify the treatments into three different groups. This is shown in Table 3.1.

The Group I processes were characterized as producing surfaces which have thin oxide layers with little macro-roughness, i.e., unevenness with characteristic bumps or jagged edges, with dimensions on the order of 1.0 μm or greater, or micro-roughness, i.e., fine structure with dimensions of 0.1 μm or less. The Group II processes all resulted in surfaces which had a high degree of macro-roughness. The Group III processes produced surfaces which were characterized by a primarily micro-rough porous oxide. This oxide is relatively thick in comparison to the oxide layers of Group I and Group II processes. Ditchek, Breen and Venable [18] described the surface morphology of the various titanium prebond treatments.

Figure 3.1 shows a collage of electron micrographs of the surface of a phosphate-fluoride treated titanium specimen. The phosphate-fluoride and the modified phosphate-fluoride treated surfaces appear very similar, there is little macro-roughness. At higher magnification fine differences can be detected as is shown in Figure 3.2A and 3.2B. For these treatments the resulting surfaces are relatively smooth with no distinguishing features. The Group II processes all produced a large degree of macro-roughness.

Table 3.1 Classification of Titanium Prebond Treatments by Surface Morphology [15]

Process	Oxide Thickness (Å)	Group Number	Comments
Phosphate-fluoride	200	I	F present
Phosphate-fluoride, stabilized	80	I	F present
Dapcotreat	60	II	No apparent fine structure, Cr present
Dry hone/Pasa-Jell 107	100–200	II	Deformed surface with embedded Al_2O_3 and F present
Liquid hone/Pasa-Jell 107	200	II	Embedded alumina, F and Cr present
Turco 5578	175	II	Fe present
Chromic acid anodize	400–800*	III	Porous oxide, F present
Alkaline peroxide	450–1,350**	III	Porous oxide

*Thickness dependent upon voltage applied, 5 and 10 volts respectively.
**Thickness dependent upon chemistry, temperature and time.

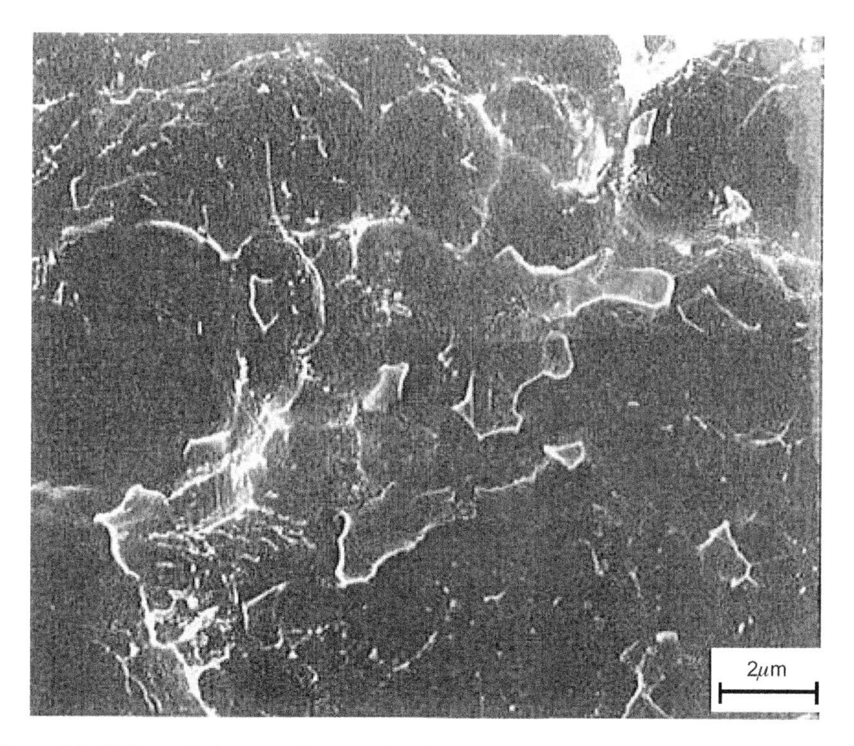

Figure 3.1 Collage of electron micrographs of the surface of phosphate-fluoride treated titanium.
Source: Courtesy of Martin Marietta Labs.

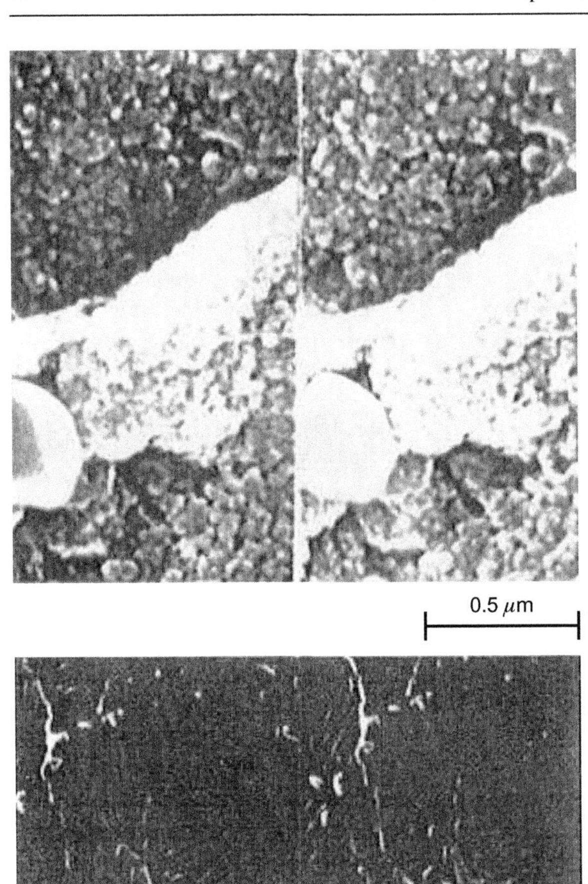

Figure 3.2 Stereo electron micrographs of titanium adherends. (A) Treated by the phosphate-fluoride process. (B) Treated by the modified phosphate-fluoride process. *Source:* Courtesy of Martin Marietta Labs.

Low magnification micrographs of the Dapcotreat and the Turco 5578 processes are shown in Figures 3.3 and 3.4 respectively. Both treatments result in protrusions that extend several microns above the surface. These protrusions, however, are reported as characteristics of the titanium substrate and not of the oxide layer. At higher magnifications there are differences between the surfaces produced by the three treatments in this group. The Daprotreat treated adherends had no

Figure 3.3 Collage of electron micrographs of the surface of the Dapcotreat treated titanium.
Source: Courtesy of Martin Marietta Labs.

distinguishable fine structure while the thicker Turco 5578 and the liquid hone/ Pasa-Jell 107 treated surfaces did, as shown in Figures 3.5 and 3.6 respectively.

The authors speculated that the macro-roughness of Group II surfaces should produce better adhesion than would the relatively smooth surfaces of Group I processes. The Group II dry hone/Pasa-Jell 107 process produces a surface (Figure 3.7) which is quite different from the other surfaces in this group. The dry hone abrasion step of the process appears to result in a deformed and fragmented

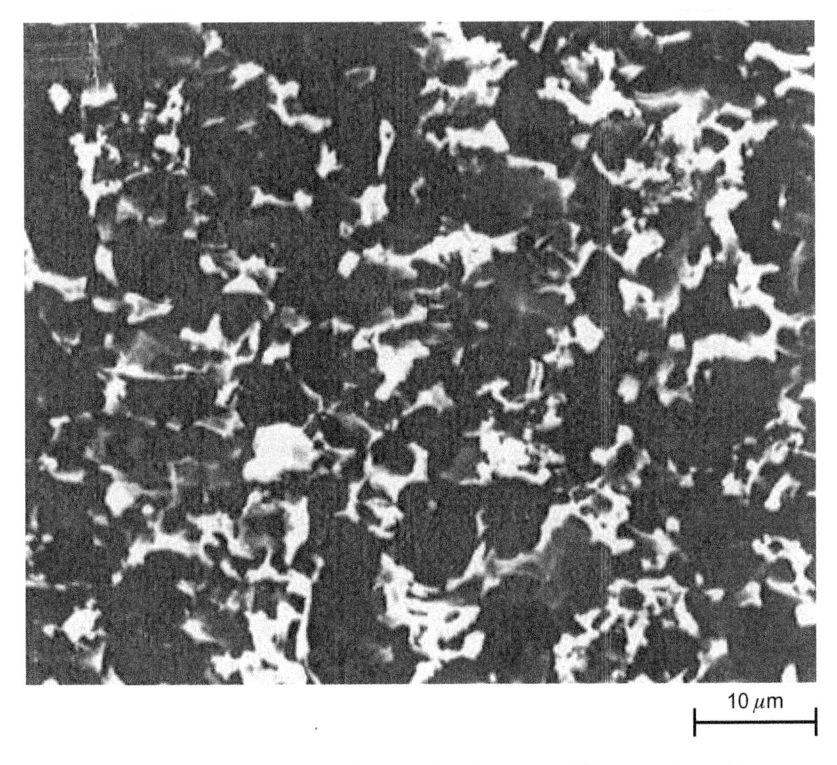

Figure 3.4 Low magnification stereomicrographs of a Turco 5578 treated titanium adherend.
Source: Courtesy of Martin Marietta Labs.

surface with a great deal of macroscopic roughness. The abrasion step also spreads aluminum oxide over the surface. Group III processes include the chromic acid anodize and the alkaline peroxide pretreatments. These two processes are characterized by thick, porous oxide surface layers.

At low magnifications the chromic acid anodized surface appears smooth, as can be seen in Figure 3.8. At higher magnifications, however, a porous structure is observed with a cell size of 300 Angstroms and protrusions which extend 300 Angstroms above the cell. However, the porous structure is not uniform all over the surface; varying amounts of the surface are actually smooth as seen in Figure 3.9A. Large amounts of smooth surface would not be desirable for good adhesion. Figure 3.9B shows the surface of the titanium adherend when treated by the alkaline peroxide process. The oxide layer was 1,350 Angstroms thick. Unlike the anodize process, however, the alkaline peroxide process produced an oxide which has no protruding whiskers.

The same processes were evaluated for their bondability. Brown [23] used the wedge test in accordance with ASTM D3762, and Wegman and Levi [24] used the stress durability test in accordance with ASTM D 2919. The data obtained were further evaluated and compared with the surface characterization groupings

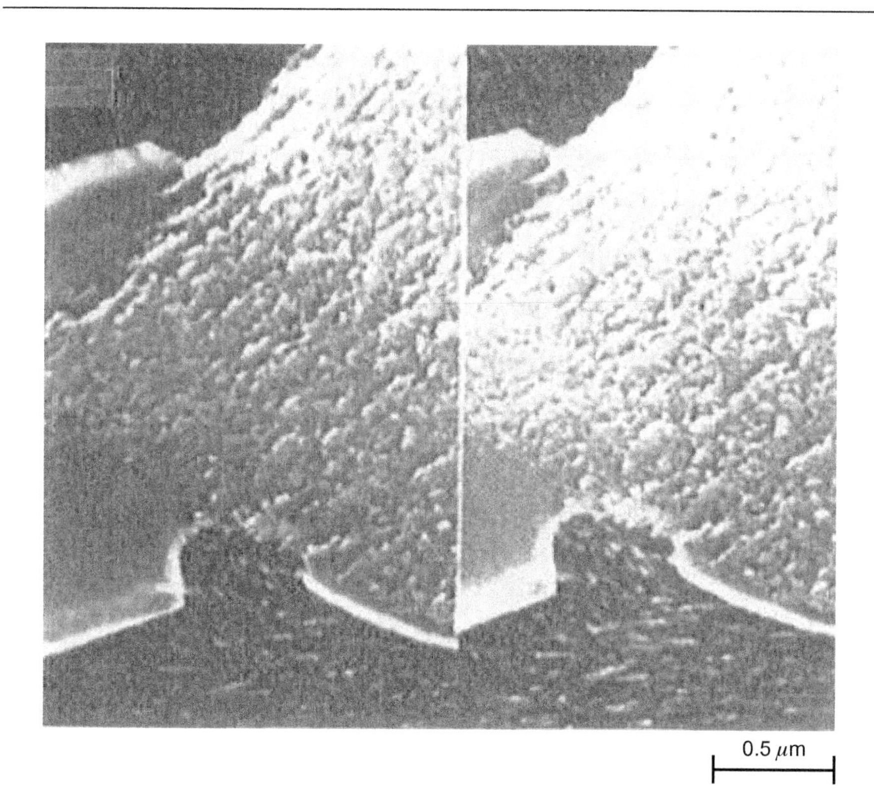

Figure 3.5 High magnification stereomicrographs of a Turco 5578 treated titanium adherend.
Source: Courtesy of Martin Marietta Labs.

discussed earlier. This work was reported by Wegman et al. [25]. The results of this comparison are shown in Table 3.2. Other evaluations of some of these same methods are given in the literature [27–31].

3.1.1 AC-130 Prebond Surface Treatment

The AC-130 surface prebond surface treatment is a Boeing Company Licensed Product Under Boegel-EPH provided by Advanced Chemistry and Technology, Inc. and registered with the U.S. Patent and Trademark Office.

The process is known as sol–gel process and was developed by the Boeing Company as a surface pretreatment for aluminum, titanium, and steel for the repair of aircraft structures at depot and field as well as OEM levels. The process was evaluated by the U.S. Air Force in conjunction with the U.S. Navy, U.S. Army, and the Boeing Company (see references 37 and 38 in Chapter 2).

The process is described as a high-performance surface preparation for adhesive bonding of aluminum alloys, steel, titanium, and composites (see reference 39 in Chapter 2).

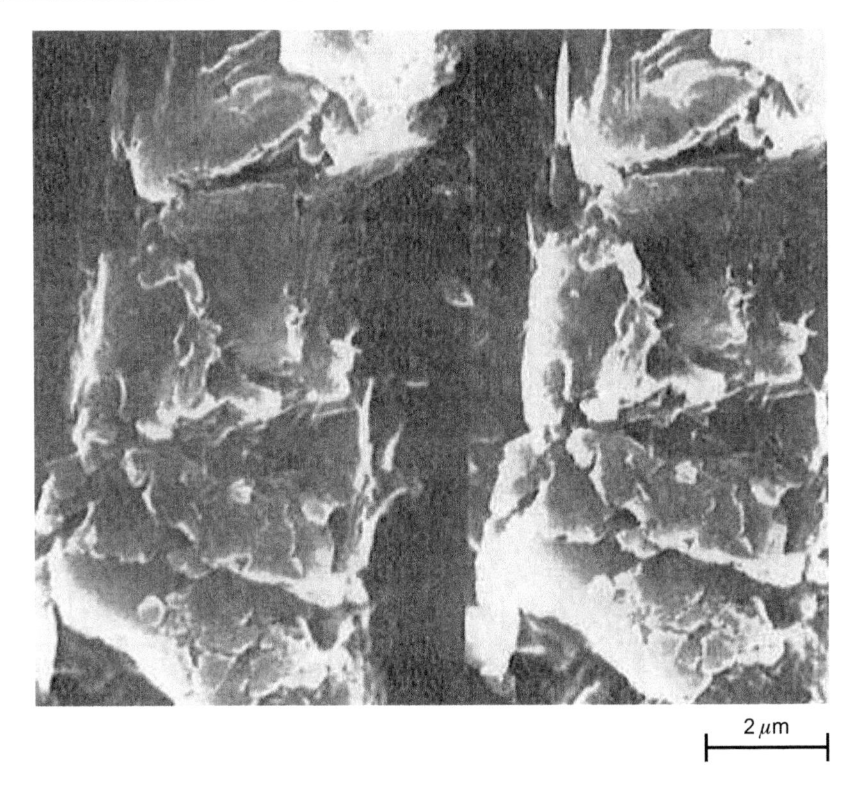

Figure 3.6 Stereo electron micrograph showing the micro-roughness characteristics of the liquid hone/Pasa-Jell 107 treated titanium adherend.
Source: Courtesy of Martin Marietta Labs.

The sol–gel process AC-130 promotes adhesion as a result of the chemical interaction at the interfaces between the adherend and the adhesive or primer.

The data have shown comparable wedge crack exposure for AC-130 prepared titanium surfaces and surfaces prepared by the CAA process (Figure 3.10).

3.2 Preparation of Titanium by the Chromic Acid Anodize Process

---

CAUTION: THE USE OF CHROMIC ACID IS CONSIDERED A HEALTH HAZARD AND IS CONSIDERED A CARCINOGEN. CHECK WITH THE PROPER GOVERNMENTAL AGENCY BEFORE CONSIDERING THE USE OF THIS METHOD.

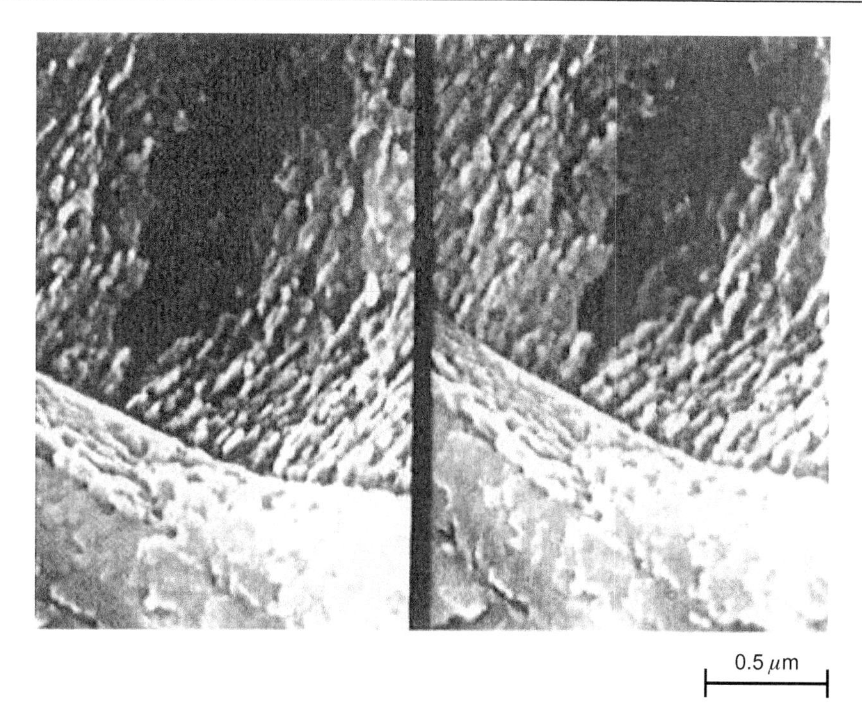

Figure 3.7 Stereo electron micrograph of the surface of the titanium adherend formed by the dry hone/Pasa-Jell **107** treatment.
Source: Courtesy of Martin Marietta Labs.

3.2.1 Apparatus

Treating Tanks

Heated tanks should be equipped with automatic temperature controls and have means for agitation to prevent local overheating of the solution. Solutions may be heated by any internal or external means that will not change their composition. Steam should not be introduced into any solution. Compressed air introduced into any solution must be filtered to remove oil and moisture.

Tank Construction

Tanks are to be made from materials that have no adverse effect on the solution used or the parts being treated. All tanks should be big enough to accept the largest part to be processed in a single treatment.

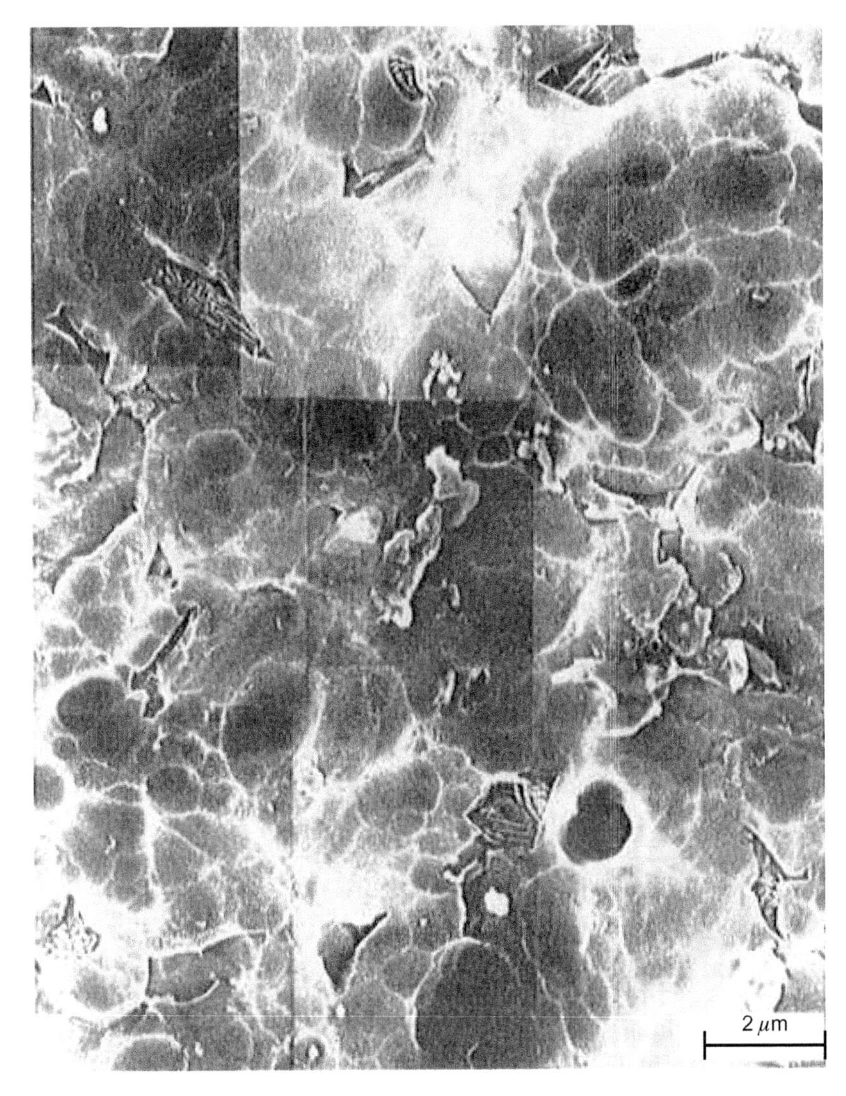

Figure 3.8 A collage of electron micrographs of the surface of a 5 volt chromic acid anodized titanium adherend.
Source: Courtesy of Martin Marietta Labs.

3.2.2 Materials

Water

Water used for makeup of processing solutions and rinsing should contain no more than 150 ppm of total solids.

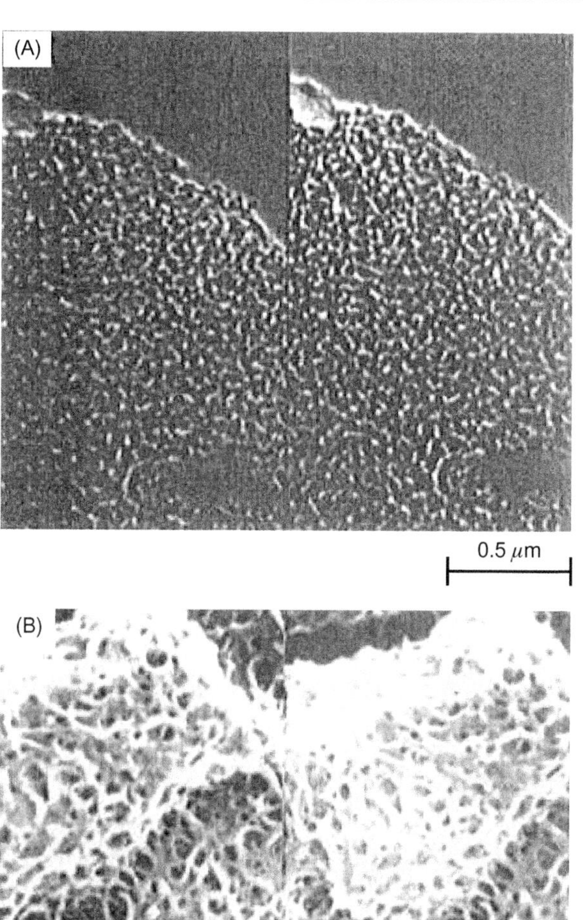

Figure 3.9 Stereo electron micrographs of a titanium adherend. (A) Treated by the 5 volt chromic acid anodize process. (B) Treated by the alkaline peroxide process. *Source:* Courtesy of Martin Marietta Labs.

Alkaline Cleaning Solution

The alkaline cleaning solution is made up of 3 to 4 pounds per gallon of Turco 5578. Protective equipment must be used; these acids are very corrosive and may be a health hazard.

Table 3.2 Comparison of Titanium Pretreatments Using the Wedge Test, the Stress Durability Test and Surface Characterization Techniques [25]

Rank	Wedge Test, 1344 hr	Stress Durability	Surface Characterization
1	CA	CA	CA, AP }
2	LP }	LP	LP }
3	AP }	AP	DA }
4	TU	DA	TU }
5	DP	TU	DP }
6	DA	DP	PF }
7	PF	PF	PFS }
8	PFS	PFS	MPFS }
9	MPFS	MPFS	

Note: } indicates the treatments were basically equivalent.
Code: CA-chromic acid anodize; LP-liquid hone/Pasa-Jell; AP-alkaline peroxide; TU-Turco 5578; DP-dry hone/Pasa-Jell; DA-Dapcotreat; PF-phosphate fluoride; PFS-phosphate fluoride, stabilized; MPFS-nitric acid/phosphate fluoride, stabilized.

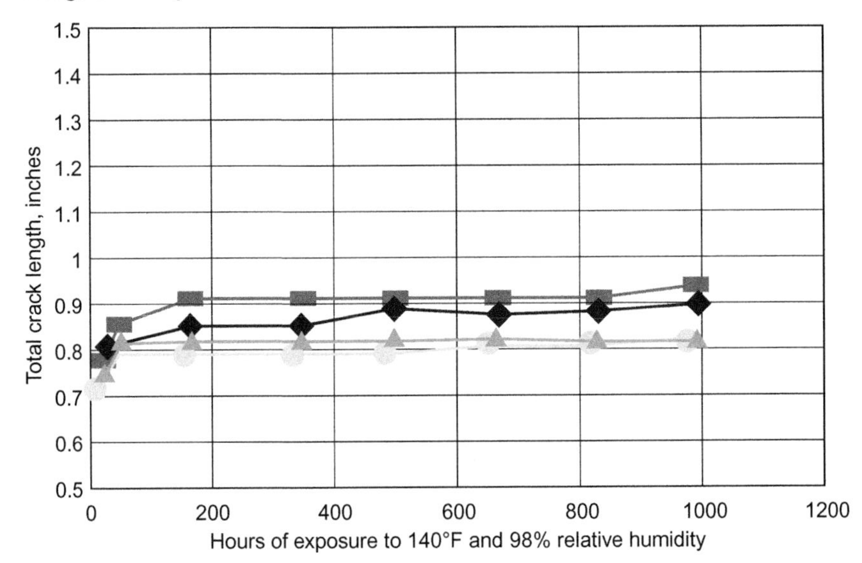

Figure 3.10 Surface treatment comparison of Ti-6Al-4V bonding. Graph courtesy of Advanced Chemistry and Technology, Inc.

Etch Solution

The etch solution is made up of 15% by volume nitric acid (70%) and 3% by volume hydrofluoric acid (50%).

Anodize Solution

The anodizing solution is made up of chromic acid (6.0 to 7.5 ounces per gallon) plus hydrofluoric acid. The hydrofluoric acid is added to obtain a current density of 1.5 amps per square foot. To increase the current density 0.5 amp per square foot add 140 ml of hydrofluoric acid for each 100 gal of solution. Current density is measured using pre-etched 6 Al, 4 V, titanium panels having an area of ½ to 1 square foot and having a surface roughness of RMS 32 or less.

3.2.3 Processing

Degreasing

Remove any excess oils and greases by any locally, safety-approved method.

Alkaline Cleaning

The parts are immersed in the alkaline cleaner at 185° to 196°F for 10 minutes.

Rinsing

The parts are rinsed in hot water (110°F minimum) for at least 5 minutes.

Etching

The parts are etched in the etch solution at room temperature for ½ to 1½ minutes.

Rinsing

The parts are rinsed in flowing water for at least 5 minutes at room temperature.

Anodizing

The parts are made the anode and anodized at 5 to 10 volts dc at 1.5 amps per square foot. The solution is operated at 60° to 80°F during the anodizing process. The voltage should be applied after the parts are immersed in the solution. The voltage is raised to the desired level within 5 minutes and maintained for 18 to 22 minutes. The anodized parts should be removed from the solution within 2 minutes after the current is shut off.

Rinsing

The anodized parts should be rinsed in cold water for 10 to 15 minutes.

Drying

Dry the parts thoroughly at 160°F maximum.

3.2.4 Restrictions

1. The color of the surface should be purple, blue-gold, blue-green or a combination of these colors. Parts with gray or spotty gray area are not acceptable.
2. Racking marks are acceptable provided there is no evidence of burning. Burns of any degree, which will be in the final part, and areas not anodized are cause for rejection.
3. Control the time between withdrawal from the processing solution and rinsing so that there is no drying of the solution on the parts.
4. Apply adhesive primer within 72 hours after drying.

3.3 Preparation of Titanium by the Turco 5578 Process

3.3.1 Apparatus

Treating Tanks

Heated tanks should be equipped with automatic temperature controls and have means for agitation to prevent local overheating of the solution. Solutions may be heated by any internal or external means that will not change their composition. Steam must not be introduced into any solution. Compressed air introduced into any solution must be filtered to remove oil and moisture.

3.3.2 Tank Construction

The tanks should be big enough to accept the largest part to be processed in a single treatment. The tanks should be manufactured from, or lined with, stainless steel and have stainless steel heating coils. Type 316, 321 or 347 stainless steel is recommended. The immersion rinse tanks are equipped with a means for skimming or overflowing or both, to remove surface contamination. Deionized water rinse tanks should be fed from the top and discharged from the bottom. Rinse tanks should be large enough to accept the total part in a single rinse.

3.3.3 Materials

Water

Deionized water is used for makeup of processing solutions and for rinsing.

Alkaline Cleaning Solution

The alkaline solution for cleaning is made up of 6 to 19 ounces per gallon of Turco 5578. Add the Turco 5578 compound to cold water with agitation. Continue agitation until the compound is completely dissolved.

Alkaline Etch Solution

The alkaline solution for etching is made up of 3 to 4 pounds per gallon of Turco 5578. Add the Turco 5578 compound to cold water with agitation. Continue agitation until the compound is completely dissolved. CAUTION: Heat is generated in the mixing process—use proper precautions; eye protection is recommended.

3.3.4 Processing

Cleaning

Immerse the parts in the alkaline cleaning solution at 165° to 175°F for 15 minutes.

Rinsing

Thoroughly rinse the parts in flowing water held at 160° to 180°F for 5 minutes.

Drying

Dry the parts in a forced air oven at 130°F.

Etching

Immerse the parts in the alkaline etch solution, which is held at 180° to 190°F, for 8 to 10 minutes.

Rinsing

Spray rinse the parts with water at 160° to 180°F for 5 to 10 minutes.

Drying

Dry the parts in a forced air oven at 130°F.

3.3.5 Restrictions

Parts to be primed should have the primer applied within 4 hours after drying. Store in kraft paper until ready for bonding.

3.4 Preparation of Titanium by the Liquid Hone/Pasa-Jell 107 Process

3.4.1 Apparatus

Treating Tanks

Heated tanks should be equipped with automatic temperature controls and have means for agitation to prevent local overheating of the solution. Solutions may be heated by any internal or external means that will not change their composition. Steam must not be introduced into any solution. Compressed air introduced into any solution is filtered to remove oil and moisture.

Tank Construction

The alkaline cleaning tank should be made from, or lined with, materials that have no adverse effect on the solution used or the parts being treated. The tank should be big enough to accept the largest part to be processed in a single treatment.

The Pasa-Jell treating tank is lined with either polyethylene or polyvinyl chloride. A polyvinyl chloride liner should be pretreated with a 50% nitric acid solution for one hour at room temperature to remove surface plasticizers before placing the tank in service. The tank is equipped with a continuous circulating pump for agitation, and should be big enough to accept the largest part to be processed in a single treatment.

3.4.2 Materials

Water

Deionized water is used for makeup of the process solutions and the final rinse. The total ionizable solids should not exceed 100 ppm on an in-line conductance bridge calibrated for sodium chloride.

Alkaline Cleaning Solution

The alkaline cleaning solution is made up of 2.5 to 3.0 lb per gallon of Turco alkaline rust remover.

Etch Solution

The etch solution is made up of 10% by volume Pasa-Jell 107-C7, 21% by volume nitric acid (42°Bé, Technical), 0.25 lb per gallon of final solution chromic acid flake and the remainder deionized water. Fill the tank half full of deionized water, add the chromic acid flake and agitate slowly. With continuous slow agitation add the nitric acid and the Pasa-Jell 107-C7. Add the remaining deionized water.

Liquid Hone

Liquid honing is done with a 220 grit aluminum oxide slurry.

3.4.3 Processing

Cleaning

The parts are cleaned with methyl ethyl ketone to remove grease or oil contamination. 1,1,1-Trichloroethane may be used in place of the methyl ethyl ketone provided that the parts are to be alkaline cleaned.

Liquid Hone

Liquid hone the faying surface of the parts with a slurry of aluminum oxide using a pressure of 60 to 80 psi.

Alkaline Cleaning

The parts should be immersed in the alkaline cleaning solution at 200° to 212°F for 20 to 30 minutes (see Restriction 5).

Rinse

Rinse the parts in an air agitated tap water or spray rinse for 2 to 4 minutes (see Restriction 4).

Etch

The parts are immersed in the etch solution at a temperature not to exceed 100°F for a period of 15 to 20 minutes.

Rinse

Rinse the parts with room temperature tap water (see Restriction 7).

Spray Rinse

The parts are spray rinsed with deionized water at room temperature for 2 to 4 minutes. Inspect for water breaks.

Drying

Dry the parts at 100° to 150°F for 30 minutes in an air circulating oven.

3.4.4 Restrictions

1. In appearance the dry hone/Pasa-Jell process produces a coating which is continuous, smooth, uniform and, when examined visually, is free from discontinuities. The color developed on 6 Al, 4 V titanium is blackish. Colors for other alloys may vary and should be determined. Surfaces may show slight variations (marbling) due to the microstructure of the alloy.
2. This method is not recommended for use where the adherend thickness is less than 0.020 inch.
3. Alkaline cleaning should be done within 2 hours of the liquid honing.
4. Seizure of the detergent solution on the titanium surface should be avoided. Keep the parts being cleaned from drying prior to rinsing. A high pressure spray rinse immediately upon removal from the detergent tank is recommended.
5. Lower the temperature of the alkaline cleaning tank to below 175°F when the tank is not in use. This is to extend the tank life.
6. The alkaline cleaner rinse tank should be air agitated.
7. The Pasa-Jell rinse tank should be agitated by use of either pump circulation or fresh water inlet flow which provides mild agitation over the entire tank. This rinse tank is used only for Pasa-Jell rinsing.
8. The liquid honing equipment should be used only on titanium.
9. Cleaned parts should be primed within 8 hours of the completion of the drying.
10. Parts may be recleaned no more than three times.

3.5 Preparation of Titanium by the Dry Hone/Pasa-Jell 107 Process

3.5.1 Apparatus

Treating Tanks

Heated tanks should be equipped with automatic temperature controls and have means for agitation to prevent local overheating of the solution. Solutions may be heated by any internal or external means that will not change their composition. Steam should not be introduced into any solution. Compressed air introduced into any solution is filtered to remove oil and moisture.

Tank Construction

The alkaline cleaning tank is made from, or lined with, materials that have no adverse effect on the solution used or the parts being treated. The tank should be big enough to accept the largest part to be processed in a single treatment.

The Pasa-Jell treating tank should be lined with either polyethylene or polyvinyl chloride. A polyvinyl chloride liner should be pretreated with a 50% nitric acid solution for one hour at room temperature to remove surface plasticizers before placing the tank in service. The tank is equipped with a continuous circulating pump for agitation, and should be big enough to accept the largest part to be processed in a single treatment.

Immersion rinse tanks for use with deionized water should be fed from the top and discharged from the bottom. Tap water rinse tanks should have a means of skimming the top of the water. Rinse tanks should be large enough to accept the total part in a single rinse.

3.5.2 Materials

Water

Rinse water used for this process may be either deionized or tap water.

Dry Hone

Dry honing is done with 320 grit virgin aluminum oxide.

Etch Solution

The etch solution consists of Pasa-Jell 107-M.

3.5.3 Processing

Cleaning

Parts should be cleaned with methyl ethyl ketone or isopropyl alcohol.

Dry Hone

The parts are blasted with clean virgin aluminum oxide grit by holding the nozzle 6 to 8 inches from the material allowing the blast to impinge the surface at an angle of approximately 45°. Use a pressure of 20 to 35 psig for materials in the 0.025 to 0.40 inch thickness range, increase the pressure for thicker materials.

Rinse

Spray or pressure rinse with water at a temperature below 100°F.

Etch

Immerse the parts in the etch solution at a temperature not to exceed 100°F for a period of 15 minutes (see Restriction 2).

Rinse

Rinse the parts by spray or immersion using water at a temperature below 100°F (see Restriction 3). Follow by immersion in a second rinse tank containing clean, filtered water for 5 to 10 minutes at a temperature below 100°F.

Dry

Dry the parts using filtered forced air or air dry at a temperature not to exceed 100°F.

3.5.4 Restrictions

1. In appearance the dry hone/Pasa-Jell process produces a coating which is continuous, smooth, uniform and, when examined visually, is free from discontinuities. The color developed on 6 Al, 4 V titanium is blackish. Colors for other alloys may vary and should be determined. Surfaces may show slight variations (marbling) due to the microstructure of the alloy.
2. It is important that the parts being treated and the Pasa-Jell 107-M solution be held at a temperature not in excess of 100°F.
3. Chromate disposal provisions must be made for this rinse water.

3.6 Preparation of Titanium by the Alkaline Peroxide Process

3.6.1 Apparatus

Treating Tanks

Heated tanks should be equipped with automatic temperature controls and have means for agitation to prevent local overheating of the solution. Solutions may be heated by any internal or external means that will not change their composition. Steam should not be introduced into any solution. Compressed air introduced into any solution is filtered to remove oil and moisture.

Tank Construction

Tanks should be made from, or lined with, materials that have no adverse effect on the solution used or the parts being treated. All tanks should be big enough to accept the largest part to be processed in a single treatment.

Rinse Tanks

Immersion rinse tanks should be equipped with a means for feeding the deionized water from the top and discharging it from the bottom. Rinse tanks should be large enough to accept the total part in a single rinse.

3.6.2 Materials

Water

Deionized water is used for makeup of process solutions and for rinsing.

Alkaline Cleaning Solution

The alkaline cleaning solution is made up with Kelite 19 in accordance with the manufacturer's instructions.

Etching Solution

The etch solution is made up to contain a mixture of 0.4 molar hydrogen peroxide and 0.5 molar sodium hydroxide solution.

3.6.3 Processing

Degreasing

The parts are wiped with methyl ethyl ketone to remove oil and grease.

Alkaline Cleaning

The parts are immersed in the alkaline cleaning solution for 3 to 5 minutes at 140° to 160°F. The solution is air agitated during the immersion cycle.

Rinsing

The parts are immersed in a tank of moving deionized water for 2 to 5 minutes. The parts are spray rinsed for 1 to 2 minutes with deionized water. Rinsing is done at room temperature.

Etching

The parts are immersed in the etch solution for 23 to 27 minutes at a temperature of 131° to 150°F.

Rinsing

The parts are immersed in a tank of moving deionized water for 2 to 5 minutes at room temperature. The parts should then be spray rinsed for 1 to 2 minutes with room temperature deionized water.

Drying

The parts should be forced air dryed for 1 hour at 150°F.

3.6.4 Restrictions

1. The hydrogen peroxide must be replenished every 30 minutes to maintain the required etching/oxidation characteristics.
2. The thickness of the oxide layer is dependent upon the temperature of the etch solution.

3.7 Preparation of Titanium by the Stabilized Phosphate Fluoride Process

3.7.1 Apparatus

Treating Tanks

Heated tanks should be equipped with automatic temperature controls and have means for agitation to prevent local overheating of the solution. Solutions may be heated by any internal or external means that will not change their composition. Steam is not introduced into any solution. Compressed air introduced into any solution is filtered to remove oil and moisture.

Tank Construction

Tanks should be made from or lined with materials that have no adverse effect on the solution used or the parts being treated. All tanks should be big enough to accept the largest part to be processed in a single treatment.

3.7.2 Materials

Water

Water used for makeup of processing solutions and for rinsing should contain no more than 150 ppm total solids. The water used for the final hot water soak should be deionized.

Alkaline Cleaning Solution

The alkaline cleaning solution is made up with Oakite HD 126 or equivalent in accordance with the manufacturer's instructions.

Pickle Solution

The pickling solution is made up of 2 to 3 fluid ounces per gallon of 70% hydrofluoric acid, 40 to 50 fluid ounces per gallon of 70% nitric acid, and 2.5 to 3 ounces per gallon of sodium sulfate.

Phosphate Fluoride Treatment

The phosphate fluoride treatment solution is made up of 6.5 to 7 ounces per gallon of trisodium phosphate, 2.5 to 3 ounces per gallon of potassium fluoride and 2.2 to 2.5 fluid ounces per gallon of hydrofluoric acid.

3.7.3 Processing

Degreasing

Remove any excess oils and greases by any locally, safety-approved method.

Alkaline Cleaning

Immerse the parts in the alkaline cleaning solution for 5 to 15 minutes at 140° to 180°F.

Rinsing

The parts are rinsed in flowing water at room temperature for 5 to 10 minutes.

Pickling

Immerse the parts in the pickle solution for 2 to 3 minutes at room temperature.

Rinsing

The parts are rinsed in flowing water at room temperature for 5 to 10 minutes.

Phosphate Fluoride Treatment

Immerse the parts in the phosphate fluoride solution for 1.5 to 2.5 minutes at room temperature.

Rinsing

The parts are rinsed in flowing water at room temperature for 5 to 10 minutes.

Hot Water Soak

The parts are soaked for 14 to 16 minutes in deionized water at a temperature of 145° to 155°F (see Restriction 1).

Final Rinse

The parts are rinsed in water at a temperature between room temperature and 160°F for 30 seconds to 1 minute.

Drying

The parts are forced air dried at a temperature between room temperature and 160°F.

3.7.4 Restrictions

1. The water used for the hot water soak should not be used for any other purpose; this water will contain fluorides.

3.8 Preparation of Titanium Alloys by the AC-130 Sol–Gel Process

3.8.1 Materials

1. AC-130 a four-part system or AC-130-2 a two-part system. AC-Tech 77341 Anaconda Ave, Garden Grove, CA92841
2. Wipers, cheesecloth, gauze, or clean cotton cloth
3. Sanding paper/discs 180-grit, or finer aluminum oxide
4. 3 M Scotch-Brite 2 in (5 cm) or 3 in (7.6 cm) medium grit "Roloc" discs
5. Solvents
6. Bonding primer
7. Proper protective equipment, such as protective gloves, respirators, and eye protection, must be worn during these operations.

3.8.2 Facilities Controls

1. Air used in this process shall be treated and filtered so that it is free of moisture, oil, and solid particles.
2. Application of the surface preparation material and primer shall be conducted in an area provided with ventilation.
3. Grinders used shall have a rear exhaust with an attachment to deliver the exhaust away from the part surface.
4. Sanding tools shall have a random orbital movement.

3.8.3 Manufacturing Controls

1. Parts to be processed shall be protected from oil, grease, and fingerprints.
2. Mask dissimilar metals and neighboring regions where appropriate.
3. Apply bond primer within 24 hours of AC-130 application. Cool parts to room temperature.
4. Apply adhesive within 24 hours of AC-130 application.
5. Contain grit and dust residues generated during mechanical deoxidization processes.

3.8.4 Storage

1. Sol–gel kits are considered to be time- and temperature-sensitive and shall be stored in accordance with the supplier's recommendations.

3.8.5 Processing

1. **Precleaning**: Remove all foreign material materials from the surface as needed.
2. Prepare the AC-130 or 130-2 in accordance with the manufacturer's instructions provided in each kit. Scale up for the size of the part as necessary, e.g., 1 liter of solution per 0.9 m^2 (10 ft^2) to be coated. Do not treat the surface with the AC-130 or 130-2 before 30 min induction time is complete. (Induction time is defined as the time period after all the AC-130 components have been mixed, but before the mixed solution is active.) Do not treat surfaces after the 10-hour maximum pot-lifetime has expired.
3. Deoxidation of the surface may be accomplished by grit blasting, sanding, mechanical Scotch-Brite, or manual Scotch-Brite.
   a. **Grit Blasting**: Using alumina grit, grit blast an area slightly larger than the bond area. Use 2–5 atm (30–80 psig) oil-free compressed air or nitrogen. Slightly overlap the blast area with each pass across the surface until a uniform matte appearance has been achieved.
   b. **Sanding**: Using a sander or high-speed grinder connected to oil-free nitrogen or compressed air line, thoroughly abrade the surface with abrasive paper for 1–2 min over a 15 × 15 cm (6 × 6 in) section covering the entire surface uniformly, following the manufacturer's instructions.
   c. **Mechanical Scotch-Brite Abrading**: Replace the abrasive paper as described in Section 3.8.1.4 with a Scotch-Brite abrasive disc and abrade surface as described in Section 3.8.1.4
   d. **Manual Scotch-Brite**: Thoroughly abrade the surface with a very fine Scotch-Brite pad for a minimum of 1–2 min as described in Section 4.3.2. Remove loose grit residue with a clean, dry, natural bristle brush or with clean oil-free compressed air or nitrogen.
4. **Application of AC-130**: Apply AC-130 or 130-2 solution as soon as possible after completion of the deoxidization process. The time between completion of deoxidization and application of AC-130 shall not exceed 30 min. The application of the AC-130 or 130-2 may be accomplished by any of the following methods: spray application, manual application using a clean natural bristle brush or swabbing with a clean wiper, cheesecloth or gauze, or by immersion. Apply solution generously, keeping the surface continuously wet with the solution for a minimum period of one minute. Surface must not be allowed to dry and should be covered with fresh solution at least one time during the solution application period. Allow the coated surface to drain for 5–10 min. If there is any surplus AC-130 solution collected in crevices, pockets, or other contained areas, use filter compressed air to lightly blow off excess solution while leaving a wet film behind. Do not splatter this excess solution onto adjoining surfaces.
5. **Drying**: Allow the coated part to dry under ambient conditions for a minimum of 60 min. Minimize contact with the part during this time, as the coating may be easily damaged until fully cured. Exact drying time will depend on the configuration of the part and room conditions (temperature and humidity).
6. **Bonding**: Apply primer and/or adhesive within 24 hours of AC-130 or 130-2 application. Keep the surface clean during the entire operation.

3.8.6 Acceptable Results

An acceptable AC-130 or 130-2 coating is smooth and continuous without evidence of surface contamination or defects. Dark areas caused by draining of the sol–gel solutions are acceptable.

References

[1] D.R. Allen, A.S. Allen, Special report, SAMPE J. April/May (1967).
[2] D.R. Allen, A.S. Allen, Titanium metal fluoride coating process, SAMPE J. April/May (1968) 73.
[3] R.F. Wegman, M.C. Ross, S.A. Slota, E.S. Duda, Evaluation of the Adhesive Bonding Processes Used in Helicopter Manufacture, Part 1. Durability of Adhesive Bonds Obtained as a Result of Processes Used in the UH-1 Helicopter, Picatinny Arsenal Technical Report no. 4186, September 1971.
[4] R.F. Wegman, W.C. Hamilton, M.J. Bodnar, A Study of Environmental Degradation of Adhesive Bonded Titanium Structures in Army Helicopters, 4th National SAMPE Technical Conference, 17–19 October 1972, pp. 425–449.
[5] R.F. Wegman, M.J. Bodnar, Structural adhesive bonding of titanium-superior surface preparation techniques, SAMPE Q. 5 (1) (1973) 28–36.
[6] R.F. Wegman, M.J. Bodnar, Durability of Bonded Titanium Joints Increased by New Process Treatments, 18th National SAMPE Symposium, 3–5 April 1974, pp. 378–384.
[7] R.F. Wegman, M.J. Bodnar, Preproduction Evaluation of an Improved Titanium Surface Prebond Treatment, 20th National SAMPE Symposium, May 1975, pp. 865–869.
[8] M.C. Ross, R.F. Wegman, M.J. Bodnar, W.C. Tanner, Effect of surface exposure time on bonding of commercially pure titanium alloy, SAMPE J. 11 (4) (1975) 4–6.
[9] M.C. Ross, R.F. Wegman, M.J. Bodnar, W.C. Tanner, Effect of surface exposure time on bonding of 6 aluminum, 4 vanadium titanium alloy, SAMPE J. 12 (1) (1976) 12–13.
[10] J.A. Marceau, Y. Moji, S.C. McMillan, A Wedge Test for Evaluating Adhesive Bonded Surface Durability, 21st National SAMPE Symposium, 6–8 April 1976, pp. 332–355.
[11] Y. Moji, J.A. Marceay, U.S. Patent 3,989,876; assigned to The Boeing Company, Seattle, WA, 1976.
[12] M.C. Locke, K.M. Harriman, D.B. Arnold, Optimization of Chromic Acid–Fluoride Anodizing for Titanium Prebond Surface Treatment, 25th National SAMPE Symposium, 6–8 May 1986, pp. 1–12.
[13] J.J. Gurtowski, S.G. Hill, Environmental Exposure on Thermoplastic Adhesives, 11th National SAMPE Technical Conference, 13–15 November 1979, pp. 252–281.
[14] R.J. Kuhbander, T.J. Aponyi, Thermal 600 Adhesive Formulation Studies, 11th National SAMPE Technical Conference, 13–15 November 1979, pp. 295–309.
[15] B.M. Ditcheck, K.R. Breen, T.S. Sun, J.D. Venables, Morphology and Composition of Titanium Adherends Prepared for Adhesive Bonding, 25th National SAMPE Symposium, 6–8 May 1980, pp. 13–24.
[16] C.L. Hendricks, S.G. Hill, Evaluation of High Temperature Structural Adhesives, 25th National SAMPE Symposium, 6–8 May 1980, pp. 39–55.
[17] V.Y. Steger, Structural Adhesive Bonding Using Polyimide Resins, 12th National SAMPE Technical Conference, 7–9 October 1980, pp. 1054–1059.
[18] B.M. Ditchek, K.R. Breen, J.D. Venables, Bondability of Titanium Adherends, Report No. MML TR-17-C, Martin Marietta Laboratory, Baltimore, MD, April 1980.
[19] B.M. Ditchek, K.R. Breen, T.S. Sun, J.D. Venables, S.R. Brown, Bondability of Titanium Adherends, 12th National SAMPE Technical Conference, 7–9 October 1980, pp. 882–895.

[20] M. Natan, J.D. Venables, K.R. Breen, Effect of Moisture on Adhesively Bonded Titanium Structures, 27th National SAMPE Symposium, 4–6 May 1982, pp. 178–191.
[21] M. Natan, K.R. Breen, J.D. Venables, Bondability of Titanium Adherends, MML TR-81–42-C, Martin Marietta Laboratory, Baltimore, MD, 1981.
[22] M. Natan, J.D. Venables, Bondability of Titanium Adherends, MML TR-82-20-C, Martin Marietta Laboratory, Baltimore, MD, 1982.
[23] S.R. Brown, An Evaluation of Titanium Bonding Pretreatments with a Wedge Test Method, 27th National SAMPE Symposium, 4–6 May 1982, pp. 363–376.
[24] R.F. Wegman, D.W. Levi, Evaluation of Titanium Prebond Treatments by Stress Durability Testing, 27th National SAMPE Symposium, 4–6 May 1982, pp. 440–452.
[25] R.F. Wegman, E.A. Garnis, E.S. Duda, K.M. Adelson, D.W. Levi, Effect of Titanium Surface Pretreatments on the Durability of Adhesive Bonded Joints, Parts I and II, Technical Report, ARSCD-TR-83005, US Army ARADCOM, Dover, NJ, 1983.
[26] M. Natan, J.D. Venables, The stability of anodized titanium surfaces, J. Adhes. 15 (2) (1983) 125–136.
[27] W. Chen, R. Siriwardane, J.P. Wightman, Spectroscopic Characterization of Titanium 6–4 Surfaces After Chemical Pretreatments, 12th National SAMPE Technical Conference, 7–9 October 1980, pp. 896–908.
[28] A.K. Rogers, K.E. Weber, S.D. Hoffer, The Alkaline Peroxide Prebond Surface Treatment for Titanium: The Development of a Production Process, SAMPE Quarterly 13(40) 1982, pp. 13–18, and 27th National SAMPE Symposium, 4–6 May 1982, pp. 63–72.
[29] J.P. Wightman, The application of surface analysis to polymer/metal adhesion, SAMPE Q. 13 (1) (1981) 1–8.
[30] A. Mahoon, Improve Surface Pretreatments for Adhesive Bonding of Titanium Alloys, 27th National SAMPE Symposium, 4–6 May 1982, pp. 150–162.
[31] D.D. Peters, E.A. Ledbury, C.L. Hendricks, A.G. Miller, Surface Characterization and Failure Analysis of Thermally Aged, Polyimide Bonded Titanium, 27th National SAMPE Symposium, 4–6 May 1982, pp. 940–953.

4 Steel and Stainless Steel

4.1 Introduction

The steels of interest in this chapter are the carbon steels, the low alloy steels and the stainless or corrosion resistant steels. Snogren [1] states that the important factors to be considered when bonding to steel are cleanliness and descaling or rust and oxide removal, and in the case of stainless steels, passivation. There are numerous reported surface preparations for carbon, low alloy and stainless steels in the literature. Most of the published information [2–6] presents the basic process without any guide to the selection of the process for a particular application nor the selection of the process in relation to the resulting bond durability. Generally the published reports do not include all of the processing parameters required to put the process into production. Snogren [1] presented a summary of performance data on various processes used for preparing steels for adhesion. This summary showed that good initial bond strengths are obtainable with a variety of surface preparations. However, the long term reliability will vary due to corrosion or oxidation of the surface at the edge and eventually under the adhesive itself.

4.2 Carbon and Alloy Steels

In 1960 Tanner and Wegman [7] evaluated the sodium dichromate-sulfuric acid etch which was commonly used for the preparation of aluminum, as a method of preparing carbon steel for adhesive bonding. They reported that extra steps were required to remove the black residue which had formed, but that good bonds could be accomplished with a room temperature curing system. In 1965 Wegman and O'Brien [8] attempted to use the sodium dichromate-sulfuric acid etch to treat carbon steel prior to bonding with an elevated temperature (150°C) curing adhesive system. During this investigation it was reported that it was necessary to heat the treated steel to at least 175°C prior to bonding to drive out the hydrogen which had been absorbed into the steel during the treatment. If this was not done prior to curing, the resulting bond line became porous with many small bubbles or voids.

In 1970 Shields [9] discussed a phosphoric acid-alcohol treatment for steel. The phosphoric acid-alcohol treatment was further investigated by Devine [10] who reported that very good bond strengths and good durability were obtained with this process. Russell et al. [11] followed up on the earlier work of Devine and reported that the surface of the steel after treatment with the phosphoric

Surface Preparation Techniques for Adhesive Bonding.

acid-alcohol process was "strikingly similar to the surfaces produced by the FPL etch on 2024 aluminum alloy." Rosty et al. [12] showed that the alcohol could be replaced by materials which were safer for use in large quantities. This would be necessary if this process were put into production. The materials which were evaluated as replacements for the alcohol included glycerin, water and propylene glycol. Both initial bond strength and bond durability values were reported.

Trawinski [13] points out that although chemical etchants improve the durability of bonds to steel when compared to mechanical surface preparations, two major disadvantages exist. These disadvantages are the high operating temperatures required, often as high as 150°F, and the fact that many of the etchants contain substantial amounts of hexavalent chromium. Hexavalent chromium is a carcinogenic material and has become an environmental concern throughout the industry. Trawinski introduced a chemical etchant for steel which operates at 70° to 80°F and is "environmentally acceptable and produces a clean, microrough morphology" which has been shown to produce strong durable bonded joints for aluminum and titanium. The etchant contains nitric and phosphoric acids and produces a smut-free surface to which good adhesive bonds can be formed. The author also states that the surface obtained is receptive to further treatment such as chemical conversion coating. The author presents data for joints made with the nitric-phosphoric acid etch compared to grit blasted steel. A comparison is shown in Tables 4.1 and 4.2.

Table 4.1 Wedge Test Results for Nitric-Phosphoric Acid Etch vs Grit Blast [13]

Surface Preparation	Crack Length (inches)				
	Initial	1 Hour	24 Hour	96 Hour	168 Hour
Unprimed grit blast	1.52	2.49	2.69	2.90	3.27
Unprimed nitric phosphoric	1.67	1.89	2.01	2.25	2.31
Primed grit blast	1.40	1.95	2.15	2.30	2.52
Primed nitric phosphoric	1.63	1.71	1.73	1.80	1.87

Table 4.2 Unstressed Durability Test Results for Nitric-Phosphoric Acid Etch vs Grit Blast [13]

95°F, 5% Salt Spray	Primed Grit-Blasted		Primed Nitric-Phosphoric	
	Shear Strength (psi)	Percent Change	Shear Strength (psi)	Percent Change
No exposure	3,512	0.0	3,898	0.0
24 hour exposure	3,447	1.8	3,861	0.9
48 hour exposure	2,820	19.7	3,718	4.6
168 hour exposure	2,524	28.1	3,388	13.1

4.3 Bonding to Oily Steel

In 1984, Rosty et al. [12] showed that good bonds to steel could be obtained regardless of the surface preparation method used, provided that the adhesive system used was one that was cured at a temperature of at least 121°C (250°F). This is illustrated in Figure 4.1. The authors further reported that chemical cleaning of steel adherends was not necessary if heavy corrosion products were not present, and that it was even possible to bond to steel which contained a thin coat of oil without cleaning the steel prior to bonding. The key to successful bonding to oily steel, or to any steel without precleaning, is the elevated temperature cure, which must be above 121°C. Bowen and Volkmann [14] also obtained satisfactory bonds to oily steel. They used a number of epoxy and urethane systems. Their cure schedules were 16 hours at 25°C followed by a postcure of 30 minutes at 135°C (268°F).

4.3.1 *Preparation for Bonding to Oily Steel*

When bonding to oily steel the surface of the steel should be wiped with a cloth to remove any heavy layer of the oil. Then bonding can be accomplished directly to

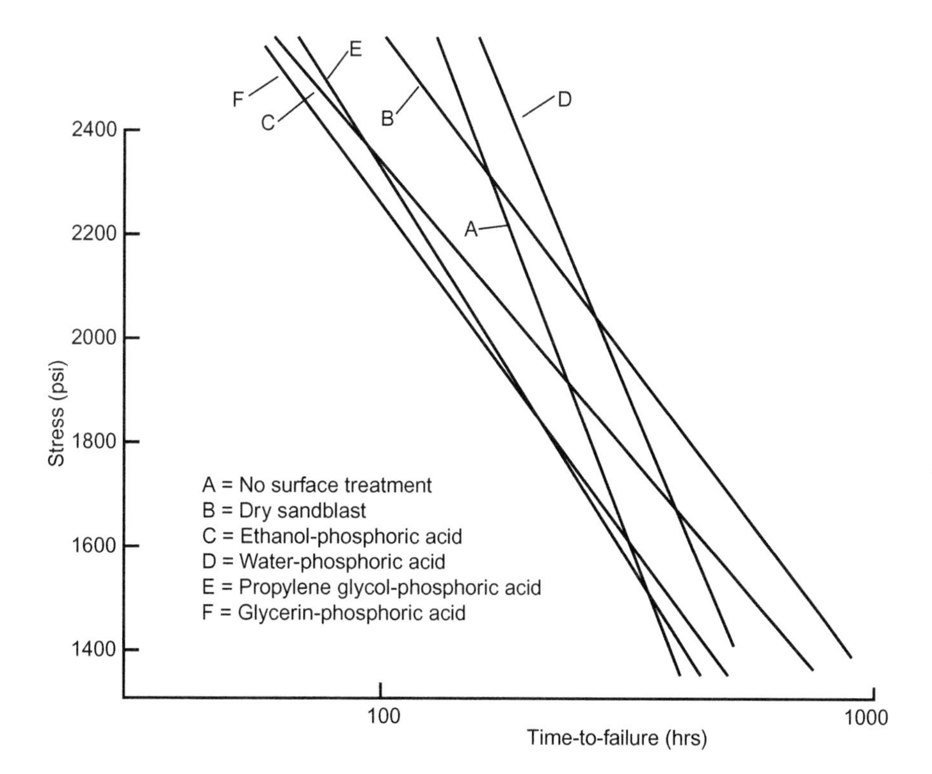

Figure 4.1 Comparison of surface treatments on steel adherends using a structural film adhesive.
Source: Courtesy of the Adhesive Section, USA ARDEC.

the surface provided that the adhesive system used is cured at a temperature above 121°C (250°F).

4.4 Conversion Coating of Steel

Corrosion resistant conversion coatings for steel are available and are used as a pretreatment for painting. These treatments have also been investigated as pretreatments for adhesive bonding. In 1976 Ross et al. [15] evaluated two phosphate coatings. These coatings were in accordance with specifications TT-C-490 and MIL-P-16232. The authors reported that the use of corrosion resistant conversion coatings would not allow the engineer to use the full potential of the structural adhesives. Most of the joint failures encountered occurred as coating failures, and the joints were adversely effected by aging at 120°F and 95% relative humidity. The conventional phosphate conversion coatings for steel did not provide adequate protection from adverse environments and were seriously affected by elevated temperature curing cycles.

Trawinski [16] also demonstrated that conventional zinc and iron phosphate conversion coatings performed very poorly in adhesion studies and that the zinc and iron phosphate crystal are loosely bonded to each other and cannot withstand applied stresses. In this study Trawinsky introduced a microcrystalline zinc phosphate conversion coating for steel. Trawinski and Miles [16] compared this new process to various chemical etchants for steel and reported that it transfers the technology for producing good durable bonds developed for aluminum and titanium to the treatment of steel. This technology involves increasing the level of surface microstructure by the application of a microcrystalline zinc phosphate conversion coating prior to adhesive bonding. The resulting surface is reportedly corrosion resistant. The authors further report that the microcrystalline zinc phosphate conversion coating surface preparation of steel will improve bond durability and is comparable to primed, chemically etched surfaces if a supported adhesive is used.

4.5 Stainless Steel

Stainless steels are steels which contain 11% or more chromium as the primary alloying ingredient. These steels are generally considered to be corrosion resistant, hence the name stainless. There are a wide variety of stainless steels. In general, the methods of preparation to be discussed will pertain to the 300 series stainless steels. The corrosion resistance of stainless steel is accomplished by passivation. Passivation is defined as making a surface inactive, or less reactive, by means of a chemical treatment. Snogren [17] states that "the passivation of stainless steel provides a chemically clean surface and aids in the formation of a thin, continuous, transparent surface film essential for good corrosion resistance." Passivation film formation occurs naturally on clean surfaces when they are exposed to the atmosphere. However, this film formation can be accomplished by immersion in certain acids or mixtures of acids and oxidizing agents. Therefore, in addition to providing corrosion resistance, these films are receptive to good adhesion. It is for this reason that various surface treatments for stainless steel have been developed using oxidizing solutions which generally are acidic.

Some of the early surface treatments for stainless steel were evaluated by Muchnick [18]. These included alkaline cleaning, alkaline cleaning followed by chromic acid or sulfuric acid and sodium dichromate etching, sulfuric acid pickle followed by nitric acid passivation, hydrochloric acid, hydrofluoric acid, nitric acid, phosphoric acid and acetic acid treatments. Wegman [19] evaluated the alkaline cleaning-chromic acid treatment and found that good bonds were attainable. There were, however, indications of surface degradation if the bonds were not made within 24 hours of preparation.

One of the methods that has been used for many years for treating 300 stainless steels for bonding is the use of strong sulfuric acid followed by immersion in a solution of sulfuric acid and sodium dichromate. Rogers [20] pointed out that the strong acid solutions produce a characteristic surface etching with a related weight loss. Thus, when thin gauge materials are to be bonded, other methods may be required. Rogers described a sulfuric acid-sodium bisulfate etch which is reported to be suitable for treating 300 series stainless steels. He showed that standard production tanks lined with chemical lead and equipped with lead heating coils can cause a problem. When as little as 5 ppm lead was dissolved in the solution the etching activity of the solution changed. The activity of the solution appeared to decrease and nonuniform etching occurred. To prevent this, a plastic liner and plastic heating coils were installed in the tank. The sulfuric acid-sodium bisulfate etch process was further investigated by Pocius et al. [21]. This modification of the sulfuric acid-sodium bisulfate etch is reported to reduce the likelihood of hydrogen embrittlement of the metal.

4.6 Specialty Stainless Steels

AM 355 stainless steel is a specialty stainless steel which has some unique properties. This steel can be cold worked to high strength, is relatively easy to form and has good corrosion resistance. Some early work on the attempts to find improved processing methods for the preparation of this steel for adhesive bonding was done under a contract sponsored by the Department of the Army and reported by Smith and Haak [22]. This work was followed up in a program conducted by Hirko et al. [23], which showed that a hydrochloric acid-ferric chloride solution resulted in a surface to which satisfactory bonds could be made while not affecting the fatigue properties of the metal nor causing any hydrogen embrittlement.

The process is known as sol–gel process and was developed by the Boeing Company as a surface pretreatment for aluminum, titanium, and steel for the repair of aircraft structures at depot and field as well as OEM levels. The process was evaluated by the U.S. Air Force in conjunction with the U.S. Navy, U.S. Army, and the Boeing Company (see references 37 and 38 in Chapter 2). The process is described as a high-performance surface preparation for adhesive bonding of aluminum alloys, steel, titanium, and composites (see reference 39 in Chapter 2).

AC-130 is a sol–gel process that promotes adhesion as a result of the chemical interaction at the interfaces between the adherend and the adhesive or primer. The AC-130 surface prebond surface treatment is a Boeing Company Licensed Product Under Boegel-EPH provided by Advanced Chemistry and Technology, Inc. and registered with the U.S. Patent and Trademark Office.

The pH or precipitation hardened stainless steels are another class of specialty steels which have been the subject of interest for use in adhesive bonded structures. However, there has been very little published on this type of material. The most widely used surface preparation process appears to be abrasive blasting.

4.7 Preparation of AM355 Stainless Steel by the AC-130 Sol–Gel Process

4.7.1 Materials

1. AC-130 a four-part system or AC-130-2 a two-part system. AC-Tech 77341 Anaconda Ave, Garden Grove, CA92841
2. Wipers, cheesecloth, gauze, or clean cotton cloth
3. Sanding paper/discs 180-grit, or finer aluminum oxide
4. 3 M Scotch-Brite 2 in (5 cm) or 3 in (7.6 cm) medium grit "Roloc" discs
5. Solvents
6. Bonding primer
7. Proper protective equipment, such as protective gloves, respirators, and eye protection, must be worn during these operations.

4.7.2 Facilities Controls

1. Air used in this process shall be treated and filtered so that it is free of moisture, oil, and solid particles.
2. Application of the surface preparation material and primer shall be conducted in an area provided with ventilation.
3. Grinders used shall have a rear exhaust with an attachment to deliver the exhaust away from the part surface.
4. Sanding tools shall have a random orbital movement.

4.7.3 Manufacturing Controls

1. Parts to be processed shall be protected from oil, grease, and fingerprints.
2. Mask dissimilar metals and neighboring regions where appropriate.
3. Apply bond primer within 24 hours of AC-130 application. Cool parts to room temperature.

4. Apply adhesive within 24 hours of AC-130 application.
5. Contain grit and dust residues generated during mechanical deoxidization processes.

4.7.4 Storage

1. Sol–gel kits are considered to be time- and temperature-sensitive and shall be stored in accordance with the supplier's recommendations.

4.7.5 Processing

1. **Precleaning**: Remove all foreign materials from the surface as needed.
2. Prepare the AC-130 or 130-2 in accordance with the manufacturer's instructions provided in each kit. Scale up for the size of the part as necessary, e.g., 1 liter of solution per 0.9 m^2 (10 ft^2) to be coated. Do not treat the surface with the AC-130 or 130-2 before 30 min induction time is complete. (Induction time is defined as the time period after all the AC-130 components have been mixed, but before the mixed solution is active.) Do not treat surfaces after the 10-hour maximum pot-lifetime has expired.
3. Deoxidation of the surface may be accomplished by grit blasting, sanding, mechanical Scotch-Brite, or manual Scotch-Brite.
   a. **Grit Blasting**: Using alumina grit, grit blast an area slightly larger than the bond area. Use 2–5 atm (30–80 psig) oil-free compressed air or nitrogen. Slightly overlap the blast area with each pass across the surface until a uniform matte appearance has been achieved.
   b. **Sanding**: Using a sander or high-speed grinder connected to oil-free nitrogen or compressed air line, thoroughly abrade the surface with abrasive paper for 1–2 min over a 15×15 cm (6×6 in) section covering the entire surface uniformly, following the manufacturer's instructions.
   c. **Mechanical Scotch-Brite Abrading**: Replace the abrasive paper as described in Section 4.7.1.3 with a Scotch-Brite abrasive disc and abrade surface as described in Section 4.7.1.4
   d. **Manual Scotch-Brite**: Thoroughly abrade the surface with a very fine Scotch-Brite pad for a minimum of 1–2 min as described in Section 4.7.1. Remove loose grit residue with a clean, dry, natural bristle brush or with clean oil-free compressed air or nitrogen.
4. **Application of AC-130**: Apply AC-130 or 130-2 solution as soon as possible after completion of the deoxidization process. The time between completion of deoxidization and application of AC-130 shall not exceed 30 min. The application of the AC-130 or 130-2 may be accomplished by any of the following methods: spray application, manual application using a clean natural bristle brush or swabbing with a clean wiper, cheesecloth or gauze, or by immersion. Apply solution generously, keeping the surface continuously wet with the solution for a minimum period of one minute. Surface must not be allowed to dry and should be covered with fresh solution at least one time during the solution application period. Allow the coated surface to drain for 5–10 min. If there is any surplus AC-130 solution collected in crevices, pockets, or other contained areas, use filter compressed air to lightly blow off excess solution while leaving a wet film behind. Do not splatter this excess onto adjoining surfaces.
5. **Drying**: Allow the coated part to dry under ambient conditions for a minimum of 60 min. Minimize contact with the part during this time, as the coating may be easily damaged

until fully cured. Exact drying time will depend on the configuration of the part and room conditions (temperature and humidity).

6. **Bonding**: Apply primer and/or adhesive within 24 hours of AC-130 or 130-2 application. Keep the surface clean during the entire operation.

4.7.6 Acceptable Results

An acceptable AC-130 or 130-2 coating is smooth and continuous without evidence of surface contamination or defects. Dark areas caused by draining of the sol–gel solutions are acceptable.

4.8 Preparation of Steel by the Phosphoric Acid-Alcohol Process

4.8.1 Apparatus

Treating Tanks

Heated tanks should be equipped with automatic temperature controls and have means for agitation to prevent local overheating of the solution. Solutions may be heated by any internal or external means that will not change their composition. Steam should not be introduced into any solution. Compressed air introduced into any solution must be filtered to remove oil and moisture.

Tank Construction

Tanks should be made from, or lined with, materials that have no adverse effect on the solution used or the parts being treated. All tanks should be big enough to accept the largest part to be processed in a single treatment.

Rinse Tanks

Immersion rinse tanks are equipped with a means for feeding the deionized water from the top and discharging it from the bottom. Rinse tanks should be large enough to accept the total part in a single rinse.

4.8.2 Materials

Water

Deionized water should be used for makeup of the process solutions and for rinsing.

Grit Blast Material

The material used to grit blast the steel is 150 grit aluminum oxide powder.

Etch Solution

The etch solution consists of 2 parts by volume ethanol and 1 part by volume 85% phosphoric acid.

4.8.3 Processing

Grit Blasting

The parts are grit blasted with aluminum oxide powder using air pressure at 25 to 35 psig.

Etching

The parts are immersed in the etch solution at a temperature of 135° to 145°F for 10 to 12 minutes.

Rinsing

The parts are immersed in running deionized water for 2 to 5 minutes at room temperature, and then spray rinsed for 1 to 2 minutes with room temperature deionized water.

Drying

The parts are dried in a forced air oven at 140° to 150°F.

4.8.4 Variations

This process may be varied by replacing the etch solution with one of the following solutions. The etching temperature for the replacement etchants is 145° to 155°F with a 10 to 12 minute etching time.

Glycerin-Phosphoric Acid Etch Solution

The etch solution consists of 230 parts by volume glycerin, 20 parts by volume water and 100 parts by volume 85% phosphoric acid.

Glycol-Phosphoric Acid Etch Solution

The etch solution consists of 73 parts by volume propylene glycol and 29 parts by volume 85% phosphoric acid.

4.8.5 Restrictions

The available data for the replacement etchants are only for 4340 steel. Potential users should conduct further evaluations for the etch time and temperatures if other steels are to be used.

4.9 Preparation of Steel by the Nitric-Phosphoric Acid Process

4.9.1 Apparatus

Tank Construction

Tanks should be made from, or lined with, materials that have no adverse effect on the solution used or the parts being treated. All tanks should be big enough to accept the largest part to be processed in a single treatment.

Rinse Tanks

Immersion rinse tanks should be equipped with a means for feeding the deionized water from the top and discharging it from the bottom. Rinse tanks should be large enough to accept the total part in a single rinse.

4.9.2 Materials

Water

Deionized water is used for makeup of process solutions and for rinsing.

Etch Solution

The etch solution consists of 30 parts by volume 85% phosphoric acid, 5 parts by volume 40° Baume nitric acid, 0.01 part by volume DuPont Zomyl FSC surfactant, and 64.99 parts by volume deionized water.

4.9.3 Processing

Etching

The parts are immersed in the etch solution at room temperature for 5 to 7 minutes.

Rinsing

The parts are immersed in running deionized water at room temperature for 2 to 5 minutes and then spray rinsed with deionized water at room temperature for 1 to 3 minutes.

Drying

The parts are dried in a forced air oven at 140° to 145°F.

4.10 Preparation of Stainless Steel by the Sulfuric Acid-Sodium Dichromate Process

CAUTION: THE USE OF SODIUM DICHROMATE IS CONSIDERED A HEALTH HAZARD AND IS CONSIDERED A CARCINOGEN. CHECK WITH THE PROPER GOVERNMENTAL AGENCY BEFORE CONSIDERING THE USE OF THIS METHOD.

4.10.1 Apparatus

Treating Tanks

Heated tanks should be equipped with automatic temperature controls and have means for agitation to prevent local overheating of the solution. Solutions may be heated by any internal or external means that will not change their composition. Steam must not be introduced into any solution. Compressed air introduced into any solution must be filtered to remove oil and moisture.

Tank Construction

Tanks should be made from, or lined with materials that have no adverse effect on the solutions used or the parts being treated. All tanks should be big enough to accept the largest part to be processed in a single treatment.

Rinse Tanks

Immersion rinse tanks should be equipped with a means for feeding the deionized water from the top and discharging it from the bottom. Rinse tanks should be large enough to accept the total part in a single rinse.

4.10.2 Materials

Water

Deionized water is used for makeup of process solutions and for rinsing.

Alkaline Cleaning Solution

The alkaline cleaning solution is made up of Oakite 164 in accordance with the manufacturer's instructions.

Etch Solution

The etch solution consists of 25 to 35 parts by volume sulfuric acid and the remainder deionized water.

Passivation Solution

The passivation solution consists of 22% to 28% by weight of sulfuric acid and 2% to 3% by weight of sodium dichromate in deionized water.

4.10.3 Processing

Degreasing

If necessary remove excess oils and greases by solvent cleaning or by use of emulsion or alkaline cleaners. Emulsion or alkaline cleaners should be used in accordance with the manufacturers' recommendations.

Alkaline Cleaning

The parts are immersed in the alkaline cleaning solution for 5 to 10 minutes at 140° to 180°F. Repeat as necessary to remove soil.

Rinsing

The parts are immersed in a tank of moving deionized water for 2 to 5 minutes at room temperature and then spray rinsed at room temperature with deionized water.

Etching

The parts are immersed in the etch solution at 135° to 145°F for 4 to 8 minutes. Timing does not start until gassing is evident on the parts. The parts should be racked with stainless steel wires.

Note: In order to start the reaction the parts may be immersed in the acid solution for about 1 minute and then raised above the solution until the acid starts to react with the steel (approximately 5 minutes). The parts are then lowered into the acid for the required time. Rubbing a part with a piece of plain steel may also be used to help start the etching action.

Rinsing

The parts may either be rinsed in deionized water or placed immediately into the passivation solution.

Passivation

The parts are immersed in the passivation solution at 140° to 160°F for 1 to 5 minutes.

Rinsing

The parts are immersed in room temperature deionized water for 1 to 5 minutes.

Final Rinse

The parts are immersed in deionized water at 140° to 160°F for 1 to 3 minutes.

Drying

The parts are dried at a temperature not to exceed 160°F.

4.10.4 Restrictions

The parts are primed or bonded within 30 days after treatment.

4.11 Preparation of Stainless Steel by the Sulfuric Acid-Sodium Bisulfate Process

4.11.1 Apparatus

Treating Tanks

Heated tanks should be equipped with automatic temperature controls and have means for agitation to prevent local overheating of the solution. Solutions may be heated by any internal or external means that will not change their composition. Steam must not be introduced into any solution. Compressed air introduced into any solution must be filtered to remove oil and moisture.

Tank Construction

Tanks should be made from, or lined with, polypropylene and have polypropylene heating coils. All tanks should be big enough to accept the largest part to be processed in a single treatment.

Rinse Tanks

Immersion rinse tanks are equipped with a means for feeding the deionized water from the top and discharging it from the bottom. Rinse tanks should be large enough to accept the total part in a single rinse.

4.11.2 Materials

Water

Deionized water is used for makeup of process solutions and for rinsing.

Alkaline Cleaning Solution

The alkaline cleaning solution is made up of Oakite 164 in accordance with the manufacturer's instructions.

Etch Solution

The etch solution consists of 1 pound of sodium bisulfate in each gallon of 7.5% by volume sulfuric acid-deionized water solution.

Passivation Solution

The passivation solution consists of 3% sodium dichromate in a 25% sulfuric acid-deionized water solution.

4.11.3 Processing

Degreasing

If necessary remove excess oils and greases by solvent cleaning or by use of emulsion or alkaline cleaners. Emulsion or alkaline cleaners should be used in accordance with the manufacturers' recommendations.

Alkaline Cleaning

The parts are immersed in the alkaline cleaning solution for 5 to 10 minutes at 140° to 180°F. Repeat as necessary.

Rinsing

The parts are immersed in a tank of moving deionized water for 2 to 5 minutes, then spray rinsed at room temperatura for 1 to 2 minutes with deionized water.

Etching

The parts are immersed in the etch solution at 135° to 145°F for 9 to 10 minutes.

Rinsing

The parts are immersed in a tank of moving deionized water for 2 to 5 minutes, then spray rinsed at room temperature for 1 to 2 minutes with deionized water.

Passivation

The parts are immersed in the passivation solution at 140° to 160°F for 1 to 5 minutes.

Rinsing

The parts are immersed in room temperature deionized water for 1 to 5 minutes.

Final Rinse

The parts are immersed in deionized water at 140° to 160°F for 1 to 3 minutes.

Drying

The parts are dried at a temperature not to exceed 160°F.

References

[1] R.C. Snogren, Handbook of Surface Preparation, Palmerton Publishing Co., Inc., NY, NY, 1974.
[2] C.V. Cagle, Adhesive Bonding, McGraw-Hill Book Co., NY, NY, 1968.
[3] ASTM Method D2651, Standard Practice for Preparation of Metal Surfaces for Adhesive Bonding, American Society for Testing and Materials.
[4] R. Keith, Martin, Adhesive Bonding of Stainless Steels, Including Precipitation-Hardening Stainless Steels, NASA-SP-5085.
[5] A. Landrock, Effect of Varying Processing Parameters in Fabrication of Adhesive-Bonded Structures, Part I Annotated Bibliography, Picatinny Arsenal Technical Report January 1970.
[6] MIL-A-9067C, Adhesive Bonding Process and Inspection, Requirements for.
[7] W.C. Tanner, R.F. Wegman, Epoxidized Novolac Resins as Structural Adhesives, Picatinny Arsenal Technical Report FRL-TR-20 December 1960.
[8] R.F. Wegman, E.L. O'Brien, Response of epoxy adhesives when stressed to failure in milliseconds, J. Appl. Polym. Sci. 10 (1966) 291–301.
[9] J. Shield, Surface preparation, Chapter 6 Adhesive Handbook, December, CRC Press, 1970
[10] A.T. Devine, Adhesive Bonded Steel: Bond Durability as Related to Selected Surface Treatments, Technical Report ARLCD-TR-77027, ARRADCOM, Dover NJ, 1977.
[11] W.J. Russell, R. Rosty, S. Whalen, J. Zideck, M.J. Bodnar, Preliminary Study of Adhesive Bond Durability on 4340 Steel Substrates, Technical Report ARSCD-TR-81020. ARRADCOM, Dover, NJ, April 1981.
[12] R. Rosty, R. Wagner, R.F. Wegman, M.J. Bodnar, Structural Adhesive Bonding of Steel, 16th National SAMPE Technical Conference, 9–11 October 1984, pp. 664–655.
[13] D. Trawinski, A low temperature etchant surface preparation for steel adhesive bonding, SAMPE Q. 16 (1) (1984) 1–5.
[14] D.O. Bowen, C.L. Volkmann, Some Adhesive Polymer Variations and Effects on Structural Bonding of Oil-Treated Metal Surfaces, 17th National SAMPE Technical Conference, 22–24 October 1985, pp. 532–542.
[15] M.C. Ross, R.F. Wegman, M.J. Bodnar, W.C. Tanner, Strength of joints to selected corrosion resistant finishes, SAMPE J. 12 (5) (1976) 4–6.

[16] D.L. Trawinski, R.D. Miles, A Comparison of Conversion Coating and Chemical Etching Surface Preparations for Steel, 16th National SAMPE Technical Conference, 9–11 October 1984, pp. 633–643.
[17] Reference 1, paragraph 1.2.2 pp. 264.
[18] S.N. Muchnik, Treatment of Metal Surfaces for Adhesive Bonding, WADC Technical Report No. 55–87, Franklin Institute Laboratories for Research and Development, Philadelphia, PA, February 1955.
[19] R.F. Wegman, Effect of time elapsed between surface preparation and bonding on adhesive joint strengths, SAMPE J. 4 (1) (1968) 19–21.
[20] N.L. Rogers, Lead Interference in Bisulfate Etch Solution for Stainless Steels, 21st National SAMPE Symposium, 6–8 April 1976, pp. 608–619.
[21] A.V. Pocius, C.J. Almer, R.D. Waid, T.Y. Wilson, B.E. Davidian, Investigation of Variability in the Adhesive Bonding Characteristics of 301 Stainless Steel, 29th National SAMPE Symposium, 3–5 April 1984, pp. 249–261, and SAMPE Journal 20(6) (1984) pp. 11–16.
[22] T. Smith, R. Haak, Treatment of AM 355 Steel for Adhesive Bonding, Rockwell International Science Center August, 1979.
[23] A.G. Hirko, J.E. Schibier, L.G. Taylor, Improved Surface Preparation of AM 355 for Adhesive Bonding, 29th National SAMPE Symposium, 3–5 April 1984, pp. 282–291.

5 Copper and Copper Alloys

5.1 Introduction

Copper and its alloys, brass and bronze, can be treated to permit the formation of good adhesive bonds. Snogren [1] and Alstadt [2] report that one of the major difficulties in adhesive bonding of copper is its tendency to form brittle amine compounds with the amine curing agents frequently used to cure many adhesive systems. Rider [3], Vazirani [4], and Dahringer [5] also discuss the corrosion of amines on copper metal. Snogren further reports that the effect of the amine can be offset by using a surface treatment which will form a barrier layer on the surface of the metal and at the same time be receptive to the adhesive being used. The barrier layer can be formed by using the black-oxide process [6–8] or the chromate conversion coating process [9–11]. The etch solutions such as nitric acid bright dip [13] do not produce a barrier layer and therefore would not be recommended for use with adhesives which use active amines in their curing systems.

Abrasion treatments such as hand sanding and sandblasting produce surfaces to which good initial bonds can be obtained. This type of surface, however, does not generally result in good durable bonds.

Ross et al. [11] evaluated the ferric chloride-nitric acid process and the ferric sulfate-sulfuric acid/sodium dichromate-sulfuric acid process with three epoxy type adhesive systems. These systems included a two part polyamide-epoxy, a one part, 250°F curing epoxy paste and a 250°F curing epoxy film adhesive. The authors concluded that the ferric sulfate-sulfuric acid/sodium dichromate-sulfuric acid treated copper surfaces yielded joints which were superior in strength to those prepared using the ferric chloride-nitric acid process. When the two-part room temperature curing polyamide epoxy was used with the ferric sulfate-sulfuric acid/sodium dichromate-sulfuric acid process the parts should be bonded within 15 days after treatment. The authors also observed that when using the ferric chloride-nitric acid process the resulting surfaces appeared to be light-sensitive. The treated surfaces turned very dark when left exposed to the light for a period of 18 to 24 hours, while those same surfaces, if protected from light, did not change color. It is interesting to note that the influence of elevated temperature (120°F) and high humidity (95% RH) did not have the normal detrimental effects on the bonded joints that had been noted for other metals.

Surface Preparation Techniques for Adhesive Bonding.

5.2 Preparation of Copper and Copper Alloys by the Ferric Chloride-Nitric Acid Process

5.2.1 Apparatus

Treating Tanks

The treating tanks should be equipped with a means for agitation to maintain a uniform concentration throughout the solution. Compressed air introduced into any solution must be filtered to remove oil and moisture.

Tank Construction

The treating tanks are made from, or lined with, materials that have no adverse effect on the solution used or the parts being treated. The tanks should be big enough to accept the largest part to be processed in a single treatment.

Rinse Tanks

Immersion tanks for potable tap water are equipped with an overflow system and fed from the bottom. Immersion tanks for deionized water are equipped with a means for feeding the deionized water from the top and discharging it from the bottom. All rinse tanks should be large enough to accept the total part in a single rinse.

5.2.2 Materials

Water

Potable tap water may be used for the initial rinse of the parts after etching. Deionized water is used to make up the process solution and for the final rinse.

Etch Solution

The etch solution consists of 15 parts by weight of a 42% ferric chloride solution (using anhydrous ferric chloride), 30 parts by weight of nitric acid (specific gravity 1.5) and 197 parts by weight of deionized water. If non-anhydrous ferric chloride is used the percentage is adjusted to compensate for water of hydration.

5.2.3 Processing

Degreasing

If necessary remove excess oils and greases by solvent cleaning or by use of emulsion or alkaline cleaners. Emulsion or alkaline cleaners should be used in accordance with the manufacturers' recommendations.

Etching

The parts are immersed in the etch solution at room temperature for 1 to 2 minutes.

Rinsing

The parts are immersed in cold running water for 2 to 3 minutes, then in cold deionized water for an additional 1 to 2 minutes.

Drying

The parts should be blown dry with clean filtered air or dry nitrogen. Hot air or oven drying will cause stains on the surface.

5.2.4 Restrictions

Protect treated surfaces from excessive exposure to light until bonding is complete.

5.3 Preparation of Copper and Copper Alloys by the Ferric Sulfate-Sulfuric Acid/Sodium Dichromate-Sulfuric Acid Process

5.3.1 Apparatus

Treating Tanks

Treating tanks should be equipped with a means of agitation to maintain a uniform concentration throughout the solution. Compressed air introduced into any solution is filtered to remove oil and moisture. Heated tanks should be equipped with automatic temperature controls. Solutions may be heated by any internal or external means that will not change the concentration of the solution. Steam should not be introduced into any solution.

Tank Construction

The treating tanks should be made from, or lined with, materials that have no adverse effect on the parts being treated or on the solutions used. The tanks should be big enough to accept the largest part to be processed in a single treatment.

Rinse Tanks

Immersion rinse tanks for potable tap water should be equipped with an overflow system and fed from the bottom. Rinse tanks for deionized water should be equipped with a means for feeding from the top and discharging from the bottom. All rinse tanks should be large enough to accept the total part in a single rinse.

5.3.2 Material

Water

If necessary remove excess oils and greases by solvent cleaning or by use of emulsion or alkaline cleaners. Emulsion or alkaline cleaners should be used in accordance with the manufacturers' recommendations.

Cleaning Solution

The cleaning solution consists of 1 part by weight ferric sulfate, 0.75 part by weight of sulfuric acid (specific gravity 1.84) and 8 parts by weight deionized water.

Chromate Conversion Solution

The chromate conversion solution consists of 1 part by weight sodium dichromate, 2 parts by weight sulfuric acid (specific gravity 1.84) and 17 parts by weight of deionized water.

Ammonium Hydroxide Solution

The ammonium hydroxide solution is a concentrated (28%) solution.

5.3.3 Processing

Degreasing

If necessary remove excess oils and greases by solvent cleaning or by use of emulsion or alkaline cleaners. Emulsion or alkaline cleaners should be used in accordance with the manufacturers' recommendations.

Cleaning

The parts are immersed in the cleaning solution at 145° to 155°F for 9 to 11 minutes.

Rinsing

The parts are rinsed in cold (room temperature or below) running water for 2 to 3 minutes.

Chromate Conversion

The parts are immersed in the chromate conversion solution at room temperature until a bright surface results.

Rinse

The parts are rinsed in cold water for 2 to 3 minutes.

Ammonium Hydroxide Rinse

The parts are immersed in the ammonium hydroxide solution for 10 seconds.

Final Rinse

The parts are immersed in cold running water for 2 to 3 minutes, then in cold running deionized water for 1 to 2 minutes.

Drying

The parts should be blown dry with clean filtered air or dry nitrogen. Hot air or oven drying will cause stains to form on the surface.

5.4 Preparation of Copper and Copper Alloys by the Black Oxide Process

5.4.1 Apparatus

Treating Tanks

Heated tanks should be equipped with automatic temperature controls. Solutions may be heated by any internal or external means that will not change their composition. Steam may not be introduced into any solution. Tanks should be equipped with a means for agitation to prevent local overheating. Compressed air introduced into any solution is filtered to remove oil and moisture.

Tank Construction

The treating tanks should be made from, or lined with, a material that has no adverse effect on the solution used or the parts being treated. All tanks shall be of a size capable of accepting the largest part to be processed in a single treatment.

Rinse Tanks

Immersion tanks for potable tap water shall be equipped with an overflow system and shall be fed from the bottom. Rinse tanks for deionized water shall be equipped with a means for feeding the deionized water from the top and discharging it from the bottom. All rinse tanks shall be large enough to accept the total part in a single rinse.

5.4.2 Materials

Water

Potable tap water or deionized water may be used throughout this process.

Degreasing

If necessary remove excess oils and greases by solvent cleaning or by use of emulsion or alkaline cleaners. Emulsion or alkaline cleaners should be used in accordance with the manufacturers' recommendations.

Cleaning Solution

The cleaning solution consists of 10 parts by volume nitric acid (70%, technical) and 90 parts by volume water.

Oxidizing Solution

The oxidizing solution consists of 240 ounces Ebanol C or its equivalent with enough water to make 1 gallon of solution.

5.4.3 Processing

Degreasing

If necessary remove excess oils and greases by solvent cleaning or by use of emulsion or alkaline cleaners. Emulsion or alkaline cleaners should be used in accordance with the manufacturers' recommendations.

Cleaning

The parts are immersed in the cleaning solution for 30 seconds at room temperature.

Rinsing

The parts are rinsed in running water for 1 to 2 minutes and then transferred immediately to the oxidizing solution. DO NOT allow the parts to dry.

Oxidizing

The parts are immersed for 1 to 2 minutes in the oxidizing solution at 97° to 99°F. DO NOT BOIL the oxidizing solution.

Rinsing

The parts are rinsed in cold water for 2 to 3 minutes.

Drying

The parts should be blown dry with clean air or dry nitrogen.

5.4.4 Restrictions

Parts prepared using this process should be bonded as soon as possible but at least within the same working day as they were processed.

This process is suitable for preparing copper alloys containing over 95% copper for use with adhesives containing chlorides and for thermally bonding to polyethylene.

5.5 Brass

5.5.1 Introduction

Brass is an alloy of copper and zinc. The ratio of copper to zinc may vary to derive various properties of the alloy. Ebneajjad [12] reported that brasses may be surface treated for bonding by sandblasting or other mechanical means or by a combination of mechanical and chemical treatments.

5.5.2 Preparation of Brass by a Mechanical–Chemical Process

Processing

1. **Mechanical**: Abrasive blast, using either a dry or a wet method. Particle size is not especially critical.
2. **Rinse**: Use deionized water.
3. **Chemical**: Treat with a 5% solution of sodium dichromate in deionized water. CAUTION: Dichromates present a health hazard.
4. **Rinse**: Rinse by immersion in deionized water.
5. Dry.

5.5.3 Preparation of Brass by the Zinc Oxide, Sulfuric Acid, and Nitric Oxide Method

Material

1. **Etch Solution**: The etch solution consists of 20 pbw zinc oxide, 460 pbw sulfuric acid, conc., and 360 pbw nitric acid 67% (sp.Gr.1.410).

Processing

1. Degrease with a safety approved solvent.
2. Etch by immersion for 5 min at 20°C (68°F).
3. Rinse in water at a temperature below 65°C (149°F).
4. Re-etch in the acid solution for 5 min at 49°C (120°F).
5. Wash in water.
6. Rinse in distilled water.
7. Dry in air (temperature of washing water and drying air must not exceed 65°C (149°F))

5.6 Bronze

5.6.1 Introduction

Bronze is an alloy of copper and tin but may contain other elements such as phosphorous, manganese, aluminum, or silicon. Ebnesajjad [13] reports that bronze may be treated for bonding by the zinc oxide, sulfuric acid, and nitric acid process given for brass.

5.6.2 Preparation of Bronze by the Zinc Oxide, Sulfuric Acid, and Nitric Acid Process

Materials

1. **Etch Solution**: The etch solution consists of 20 pbw zinc oxide, 460 pbw sulfuric acid, conc., and 360 pbw nitric acid 67% (sp.Gr.1.410).

Processing

1. Degrease with a safety approved solvent.
2. Etch by immersion for 5 min at 20°C (68°F).
3. Rinse in water at a temperature below 65°C (149°F).
4. Re-etch in the acid solution for 5 min at 49°C (120°F).
5. Wash in water.
6. Rinse in distilled water.
7. Dry in air (temperature of washing water and drying air must not exceed 65°C (149°F)).

References

[1] R.C. Snogren, Handbook of Surface Preparation, Chapter 11, Palmerton Publishing Co., Inc., NY, NY.
[2] D.M. Alstadt, Some fundamental aspects of rubber-metal adhesion, Rubber World November (1955).
[3] D.K. Rider, Adhesives in Printed Circuit Applications, Symposium on Adhesives for Structural Applications, Interscience, New York, 1962, pp. 49.
[4] H.N. Vazirani, Surface preparation of copper and its alloys for adhesive bonding and organic coatings, J. Adhesion 1 (1969) 208.
[5] D.W. Dahringer, in: M.J. Bodnar (Ed.), Corrosion by Adhesives: Some Causes and Effects, Processing for Adhesive Bonded Structures, Interscience Publishers, New York, 1972, p. 371.
[6] R.C. Snogren, Selection of surface preparation processes, Part 1 process selection guide, Adhesives Age July (1969).
[7] R.C. Snogren, Selection of surface preparation processes, Part 2 appendix of surface preparation processes, Adhesives Age August (1969).

[8] ASTM Method D2651, Standard Practice for Preparation of Metal Surfaces for Adhesive Bonding, Method E, American Society for Testing and Materials, Philadelphia, PA.
[9] Ibid, Method O.
[10] M.A. Kuchner, Selecting conversion coatings for protection and appearance, Machine Design March (1970).
[11] M.C. Ross, R.F. Wegman, M.J. Bodnar, W.C. Tanner, Effect of surface exposure time on bonding of copper alloy, SAMPE J. May/June (1975) 11–13.
[12] Reference 5, Method B.
[13] S. Ebnesajjad, Surface Treatment of Materials for Adhesive Bonding, William Andrews Publishing, Norwich, NY, USA, 2006.

6 Magnesium

6.1 Introduction

Magnesium and magnesium alloys are of interest because of their light weight. Consideration of the use of magnesium and magnesium alloys must involve concern about corrosion resistance and the means of preventing corrosion. When parts are to be adhesively bonded, the means of preventing corrosion must also be compatible with the adhesive, produce a coating that is cohesively strong and adheres well to the metal surface. Studies by Eickner [1], Muchnick [2], Jackson [3], and Mulroy and Izu [4] indicate that the best adhesion in terms of bond strength and environmental resistance may be developed by use of the chromic acid or anodic processes.

In order to obtain the most durable properties in the bonded joint, the coating developed by the process should be kept as thin as possible. This results in the strongest cohesive strength of the coating, while at the same time affording the greatest possible protection to the metal. The coating produced by the chromic acid process should be 0.1 mil or less while that produced by the anodic process should be no thicker than 0.3 mil. Snogren [5] and MIL-M-45202 [6] discuss a number of processes for treating magnesium; some of these are satisfactory for paint adhesion, others are satisfactory for use with adhesives, while still others are intended for maximum corrosion protection. This discussion will only cover the four best preparations for adhesive bonding to magnesium. These include MIL-M-45202, Type 1, Class A, Grade 2; MIL-M-45202, Type 1, Class C; chromic acid hot process; and the chromic acid room temperature process.

Table 6.1 compares the shear strength data obtained when the hot chromic acid, the room temperature chromic acid and the MIL-M-45202, Type 1, Class C, Dow 17 process were used to treat four magnesium alloys bonded with two different adhesives. The data indicated that in general the hot chromic acid prepared specimens had the highest strength followed in turn by the specimens prepared by the room temperature chromic acid process, and then the Dow 17 process. These data were obtained from Reference 3. However, there were no data available on the effects of environmental exposure on the joints prepared by these processes.

Data obtained from Reference 5, comparing the effect of environmental conditioning on bonds to Dow 17 and the MIL-M-45202 Type 1, Class A, Grade 2, HAE anodized surfaces, are shown in Table 6.2. These data show that better bond strengths are obtained by the Dow 17 process treated surfaces. However, specimens prepared by both processes retained good strength after environmental conditioning. The failure of almost all specimens was predominantly failure of the anodic coating.

Surface Preparation Techniques for Adhesive Bonding.

Table 6.1 Lap Shear Strengths to Treated Magnesium Alloys [3]

Adhesive	Alloy	Lap Shear Strength, psi		
		Hot Chromic Acid	RT Chromic Acid*	Dow 17
A**	AZ-31	2,710	2,490	2,120
	ZK60-T6	2,330	2,280	2,200
	HM-21	2,460	2,420	2,340
	HK-31	2,658	2,500	1,990
B***	AZ-31	2,210	1,800	2,110
	ZK60-T6	2,170	1,830	1,850
	HM-21	2,718	2,280	2,170
	HK-31	2,700	2,200	2,080

*RT Chromic Acid—room temperature chromic acid.
**Adhesive A—Epoxy, rubber based.
***Adhesive B—Epoxy polyamide.

Table 6.2 Lap Shear Strengths to AZ 31-H24 Magnesium Alloy Treated by the Dow 17 and the HAE Anodic Processes [5]

Environmental Conditioning	Adhesive	Tensile Shear Strength at 76°F, psi	
		Dow 17	HAE Anodize
RT* control	Metlbond	2,450	1,940
150 hr @ 250°F	4021	2,410	1,620
6 months @ 120°F and 97% RH**		1,860	1,500
30 days salt water spray		1,650	1,400
RT control	FM-47	1,510	660
150 hr @ 250°F		1,480	591
6 months @ 120°F and 97% RH		1,350	565
30 day salt water spray		1,460	530
RT control	Redux 775	1,470	618
150 hr @ 250°F		1,170	570
6 months @ 120°F and 97% RH		1,260	559
30 days salt water spray		1,660	650
RT control	Epon VIII	1,640	726
150 hr @ 250°F		1,460	1,250
6 months @ 120°F and 97% RH		876	517
30 day salt water spray		1,850	1,050
RT control	AF-6	2,580	1,780
150 hr @ 250°F		2,950	1,410
6 months @ 120°F and 97% RH		2,650	1,460
30 day salt water spray		2,680	1,570

*RT is room temperature.
**RH is room humidity.

6.2 ASTM D1732, Type A, Grade 2 (ASTM D1732 Supersedes the MIL Specification)

6.2.1 Apparatus

Treating Tanks

Tanks requiring heating or cooling shall be equipped with automatic temperature controls and shall have a means for agitation to prevent local hot or cold spots in the solution. Solutions may be heated by any internal or external means that do not change their composition. Cooling may be accomplished by use of cold water tubes or refrigerant or other suitable medium. Steam shall not be introduced into any solution. Compressed air introduced into any solution shall have been filtered to remove oil and moisture.

Tank Construction

Tanks shall be made from, or lined with, materials that have no adverse effect on the solutions used or the parts being processed. All tanks should be big enough to accept the largest part to be processed in a single treatment. A minimum of 2 inches should be allowed between the sides (including coils) of the tank and the parts being processed.

Alkaline Cleaning Tank
The recommended material of construction for the alkaline cleaning tank is steel.

Anodize Tank
The recommended material of construction for the HAE anodize tank is black iron. Valves that are in contact with the anodize solution should be made of black iron or copper. Galvanized iron, brass, tin, zinc, rubber and all oxidizable materials should be avoided.

Post-Treatment Tank
The tank for the bifluoride-dichromate post-treatment solution should be lined with polyethylene, polypropylene or similar inert material.

Rinse Tanks
Immersion rinse tanks for deionized water should be fed from the top and discharged from the bottom. If skimming is required a weir connecting to the bottom drain may be used.

Racks, Clamps and Bus Bars

Backs and clamps should be made from magnesium or magnesium alloy. The bus bar used in anodizing should be copper.

Electrical System

The electrical system should be adequate to provide a current density of 18 to 20 amp/ft^2 at a voltage of 0 to 60 V ac.

6.2.2 Appearance

The anodize process produces a coating which should be continuous, smooth, uniform in appearance and, when examined visually, should be free from discontinuities such as scratches, breaks, burns and areas which are not anodized.

Color

The anodized coating on the part should be a light tan in color.

Racking Marks

Racking marks are acceptable provided there is no evidence of burning. Burns can be identified by pitted or melted surfaces. Burns of any degree that will be in the final part area are cause for rejection.

6.2.3 Materials

Solvent Cleaning

Oil and grease are removed by solvent cleaning. This may be accomplished by soaking, spraying, vapor degreasing or ultrasonic cleaning using organic solvents or emulsion cleaners which do not attack magnesium. Do not use methyl (wood) alcohol in any cleaning formulation.

Alkaline Cleaner

An alkaline cleaner recommended for steel, or one having a pH above 8, may be used. If a proprietary formulation is used, the product manufacterer's operating instructions should be followed. A satisfactory cleaner which may be used consists of 2 to 8 oz of sodium hydroxide, 1.3 oz of sodium phosphate (12 waters of hydration), 0.1 oz of wetting agent and enough water to make one gallon of solution.

Anodize Solution

The anodize solution consists of 22 oz potassium hydroxide, 4.5 oz aluminum hydroxide (see Section Alternate Ingredients), 4.5 oz potassium fluoride, anhydrous, 4.5 oz trisodium phosphate (12 waters of hydration), 2.5 oz potassium manganate (see Section Alternate Ingredients) and enough water to make one gallon of solution. The solution should be made by dissolving the ingredients in water in the order given.

Alternate Ingredients

Instead of aluminum hydroxide 1.5 oz of scrape 1100 series aluminum alloy (99% aluminum) per gallon of solution may be used. However, the aluminum must be allowed to react in a separate container with a portion of the potassium hydroxide in solution, until dissolved, before transferring to the processing tank. Any undissolved residue is separated by decantation and discarded.

CAUTION: THE REACTION OF ALUMINUM AND POTASSIUM HYDROXIDE PRODUCES HYDROGEN. THE HYDROGEN GAS MUST BE CAREFULLY DISSIPATED. KEEP FLAMES OUT OF THE AREA.

If potassium manganate is not available an equal amount of potassium permanganate may be substituted. The permanganate must be completely dissolved in hot water before adding to the processing tank.

Post-Treatment Solution

The post-treatment solution consists of 10.8 oz of ammonium bifluoride, 2.7 oz of sodium dichromate (2 waters of hydration) and enough water to make one gallon of solution.

6.2.4 Processing

Degreasing

Remove any markings by solvent wiping. Remove any oil or grease from the surface by any of the methods described in the section Alternate Ingredients.

Alkaline Cleaning

The parts are immersed in the alkaline solution described in the section Alkaline Cleaning at a temperature of 180° to 200°F for a period of 8 to 10 minutes. The parts are flushed or spray rinsed in cold water.

Anodizing

The parts are divided into groups of approximately equal surface areas. Each group serves as an electrode. The parts are immersed in the anodize solution described in Section 6.2.4. Apply an alternating current having a frequency of 60 ± 10 cycles per second across the electrodes. The current density should be increased to 18 to 20 amp/ft^2 by increasing the voltage from 0 to 60 V ac. The current density should be held for approximately 8 minutes. Depletion of the manganate is indicated by the lightening of the characteristic color of the coating. Adjustment of the aluminum and the manganate content is normally made when 140 ft^2 of magnesium per gallon of solution has been treated.

The aluminum is dissolved separately in a sufficient amount of potassium hydroxide before adding to the tank (see the section Post-Treatment Solution). Potassium manganate is first dissolved in 5% potassium hydroxide solution. If permanganate is used it shall be dissolved in hot water and twice the amount of potassium hydroxide should be added. The free alkali content is maintained at a level of 10 to 12% free potassium hydroxide. See the appendix to MIL-M-45202 for method of analysis.

Post-Treatment

The parts are immersed in the post-treatment solution (see the section Post-Treatment Solution) at room temperature for one minute. Do not rinse. Allow to dry at room temperature.

6.3 ASTM D1732, Type 1, Class C (ASTM D1732 Supersedes the MIL Specification)

6.3.1 Apparatus

Treating Tanks

Heated tanks should be equipped with automatic temperature controls and have a means for agitation to prevent local overheating of the solution. Solutions may be heated by any internal or external means that will not change their composition. Steam must not be introduced into any solution. Compressed air introduced into any solution is filtered to remove oil and moisture.

Tank Construction

Tanks should be made from, or lined with, materials that have no adverse effect on the solutions used or the parts being treated. All tanks should be big enough to accept the largest part to be processed in a single treatment. A minimum of 2 inches must be allowed between the sides (including coils) of the tank and the parts being processed.

Alkaline Cleaning Tank
The recommended material of construction for the alkaline cleaning tank and heating coils is steel.

Pickle Tank
The pickle tank should be lined with a material which will not be attacked by hydrofluoric acid.

Anodize Tank
Mild steel is the recommended material of construction for the anodize tank and heating coils. Synthetic rubber or vinyl based materials may be used to line the tank to prevent contact of the parts with the tank. Contact with the tank during processing will cause burning. Copper, nickel, lead, chromium, zinc, aluminum and

monel are materials to be avoided in construction of the tank because they are attacked by the solution.

Rinse Tanks

Immersion rinse tanks for deionized water should be fed from the top and discharged from the bottom. If skimming is required a weir may be used. Tap water rinse tanks should be fed from the bottom and discharged at the top.

Racks and Clamps

The racks and clamps should be made from magnesium or aluminum alloy 5052 and 5056. Good electrical contact with the part is required to prevent burning. If magnesium contacts are used they will have to be stripped periodically. This can be done by immersion in a 20% chromic acid solution or by mechanical means. Aluminum contacts do not generally require cleaning.

Electrical System

The electrical system is between 5 and 50 amp/ft^2 at a voltage of up to 75 V dc. The steel tank is the cathode and the part the anode. An ammeter and a rheostat must be in the circuit.

6.3.2 Appearance

The anodize process produces a coating which should be continuous, smooth, uniform in appearance and, when examined visually, free from discontinuities such as scratches, breaks, burns and areas which are not anodized.

Color

The anodized coating on the part should be light green in color and approximately 0.0003 inch thick. Heavy or dark green coatings are cohesively weak.

Racking Marks

Racking marks are acceptable provided that there is no evidence of burning. Burns can be identified by pitted or melted surfaces. Burns of any degree, which will be in the final part area, are cause for rejection.

6.3.3 Materials

Solvent Cleaning

Oil and grease are removed by solvent cleaning. This may be accomplished by soaking, spraying, vapor degreasing or ultrasonic cleaning using organic solvents or emulsion cleaners which do not attack magnesium. Do not use methyl (wood) alcohol in any cleaning formulation.

Alkaline Cleaning

An alkaline cleaner recommended for steel or one having a pH above 8 may be used. If a proprietary formulation is used, the product manufacturer's operating instructions should be followed. A satisfactory formulation which may be used consists of 2 to 8 oz sodium hydroxide, 1.3 oz sodium phosphate (12 waters of hydration), 0.1 oz of a wetting agent and enough water to make one gallon of solution.

Pickling

The magnesium-based alloys are pickled using an appropriate procedure as described in MIL-M-3171, or by immersion in a solution consisting of 38 oz of hydrofluoric acid (60%) and enough water to make one gallon of solution.

Anodize Solution

The anodizing solution consists of 40 to 60 oz/gal of ammonium bifluoride, 6.7 to 16 oz/gal of sodium dichromate (2 waters of hydration), 6.5 to 14 fluid ounces per gallon of phosphoric acid (85%) (specific gravity 1.69), and enough water to make one gallon. The solution should be prepared as follows: Heat one-half of the required water to 160°F (71°C). Add the ammonium bifluoride slowly with constant stirring. Then add the other chemicals and the rest of the water. Heat to 180°F and stir vigorously for 5 to 10 minutes.

6.3.4 Processing

Degreasing

Remove any markings by solvent wiping. Remove any visible oil or grease from the surface by any of the methods described in the section Pickling.

Alkaline Cleaning

The parts are immersed in the alkaline cleaning solution (see the section Alkaline Cleaning) at a temperature of 180° to 200°F for a period of 8 to 10 minutes. The parts are flushed or spray rinsed with cold water.

Pickling

The parts are immersed in the pickle solution (see the section Pickling) at a temperature between 70° and 90°F for 5 minutes (30 seconds for AZ 31B alloy). The parts are rinsed by immersion in cold running water or spray rinsed with cold water.

Anodizing

The parts are made the anode in the solution specified in the section Anodize Solution. The steel tank or a steel strip (if a lined tank is used) is made the cathode. Anodizing is carried out at a constant current of between 5 and 50 amp/ft^2. The voltage is continuously increased to 75 V dc to maintain the desired current density.

The time based upon 20 amp/ft^2 is 2.5 to 3 minutes. Time varies inversely with the current density, i.e., for 10 amp/ft^2 the time would be increased to 5 to 6 minutes and at 40 amp/ft^2 the time would be decreased to 1.25 to 1.5 minutes.

Approximately 20 ft^2 of surface may be treated per gallon of solution before adjustments in the bath may be required. The amount of various ingredients to be added are determined by standard analytical methods. See the Appendix to MIL-M-54202. 5.

6.3.5 Restrictions

Safety

Because of the relatively high voltages used with this process it is recommended that additional safety precautions be used. A gate might be used to prevent contact with the tank. The gate should contain an interlock which when broken would automatically shut down the anodizing current.

Alkaline Cleaner

Alkaline cleaner containing more than 2% caustic (sodium hydroxide) will etch ZK60A and ZK60B and some magnesium-lithium alloys with resulting dimensional variations.

6.4 Preparation of Magnesium and Magnesium Alloys by the Chromic Acid Treatment Processes

CAUTION: THE USE OF CHROMIC ACID IS CONSIDERED A HEALTH HAZARD AND IS CONSIDERED A CARCINOGEN. CHECK WITH THE PROPER GOVERNMENTAL AGENCY BEFORE CONSIDERING THE USE OF THIS METHOD.

6.4.1 Apparatus

Treating Tanks

Heated tanks should be equipped with automatic temperature controls and have a means for agitation to prevent local overheating of the solution. Solutions may be heated by any internal or external means that will not change their composition.

Steam should not be introduced into any solution. Compressed air introduced into any solution is filtered to remove oil and moisture.

Tank Construction

Tanks should be made from, or lined with, materials that have no adverse effect on the solutions used or the parts being treated. All tanks should be big enough to accept the largest part to be processed in a single treatment.

Alkaline Cleaning Tank

Steel is the recommended material of construction for the alkaline cleaning tank and heating coils.

Chromate Tank

The tank for the chromic acid solution should be lined with an inert liner such as rubber or polypropylene.

Rinse Tanks

Immersion rinse tanks for deionized water should be fed from the top and discharged from the bottom. If skimming is required a weir may be used.

6.4.2 Materials

Water

The water used for solution makeup and for rinses should be deionized water.

Solvent Cleaning

The parts should be vapor degreased using trichloroethylene or may be immersed, sprayed or wiped with acetone or other organic solvents which do not attack magnesium. Do not use methyl (wood) alcohol in any cleaning formulation.

Alkaline Cleaning

An alkaline cleaner with a pH of greater than 11.0 may be used. An alkaline formulation that has been found satisfactory consists of 3.0 oz of sodium metasilicate, 1.5 oz of tetrasodium pyrophosphate, 1.5 oz of sodium hydroxide, 0.5 oz of a detergent and enough water to make one gallon.

Chromic Acid

The chromic acid solution consists of 20% by weight chromium trioxide per gallon of deionized water.

Chromic Acid Alternate (Room Temperature)

The alternate chromic acid solution consists of 24 oz chromium trioxide per gallon of deionized water.

6.4.3 Processing

Degreasing

Remove any markings by solvent wiping. Remove any oil or grease from the surface by any of the methods described in the section Solvent Cleaning.

Alkaline Cleaning

The parts are immersed in the alkaline cleaner described in the section Alkaline Cleaning at 160°F for a period of 8 to 12 minutes. The parts are thoroughly rinsed in cold deionized water.

Chromic Acid Treatment

The parts are immersed in the chromic acid solution described in the section Chromic Acid for a period of 9 to 11 minutes at 150° to 160°F. The parts are thoroughly rinsed in room temperature deionized water and air dried at a temperature of up to 150°F.

Chromic Acid Treatment Alternate (Room Temperature)

The parts are immersed in the chromic acid solution described in the section Chromic Acid Alternate (Room Temperature) for a period of 2 minutes at room temperature. The parts are rinsed in deionized water at room temperature and air dried at a temperature of up to 150°F.

References

[1] H.W. Eickner, Effect of surface treatment on adhesive bonding properties of magnesium, Forest Products Laboratory Report No. 1865, June 1958.

[2] S.N. Muchnick, Treatment of Metal Surfaces for Adhesive Bonding, The Franklin Institute Laboratories for Research and Development, WADC Technical Report No. 55–87, February 1955.

[3] L.C. Jackson, Effect of Surface Preparation on Bond Strength of Magnesium, Applied polymer symposia No. 3, John Wiley and Sons, NY, NY, 1966, pp. 341–351.

[4] B.J. Mulroy Jr., Y.D. Izu, Influence of Space Environments on Adhesive, 13th National SAMPE Technical Conference, SAMPE, Covina, CA, 1981, pp. 270–279.

[5] R.C. Snogren, Chapter 8 Handbook of Surface Preparation, Palmerton Publishing Co., NY, NY, 1974, pp. 197–261

[6] ASTM E1732 Standard Practice for Preparation of Magnesium Alloy Surfaces for Painting.

7 Other Metals

7.1 Introduction

The surface preparation of the less common metals such as beryllium, cadmium, chromium, gold, nickel, silver, tin, tungsten, uranium and zinc has not been developed for large scale production processes nor have enough data been published comparing processes to back up the selection of a best process. Therefore, some of the available methods are presented in varying details. Interested potential users of these methods are cautioned that more intensive evaluations should be conducted before these methods are automatically placed into production. Some of these methods were developed many years ago and therefore extreme care should be used to check approved safety standards covering the various listed chemicals.

7.2 Beryllium

Beryllium and its alloys are useful because of their strength and light weight. However, airborne particles of beryllium and beryllium oxide are extremely toxic. Beryllium and its alloys must be treated with extreme caution when processing as they produce dust, fine chips or slivers, scale, mists, or fumes. Special equipment should be employed to collect these waste products. All waste stream products must be properly disposed of by approved means.

CAUTION: BERYLLIUM REACTS QUICKLY WITH SUCH SOLVENTS AS METHYL ALCOHOL, FREON, PERCHLOROETHYLENE AND METHYL ETHYL KETONE/FREON. IT CAN BE PITTED BY LONG EXPOSURE TO TAP WATER CONTAINING CHLORIDES OR SULFATES.

7.2.1 Preparation of Beryllium by the Sodium Hydroxide Process [1]

1. Degrease with trichloroethylene.
2. Immerse in a solution consisting of 20 to 30 pbw sodium hydroxide and 170 to 180 pbw distilled water at 68°F (20°C) for 5 to 10 minutes.
3. Rinse in distilled water after washing in tap water.
4. Oven dry at 200° to 250°F (121° to 177°C) for 10 minutes.

Beryllium has a tendency to corrode or pit when subjected to a moist atmosphere for long periods of time. MIL-HDBK-691B recommends that a procedure which produces a thin transparent coating less than 100 Angstroms thick be used to prevent

Surface Preparation Techniques for Adhesive Bonding.

corrosion. This process employs a chromate-containing solution. Check local codes on the use and handling of chromates before setting up for this process.

7.2.2 Preparation of Beryllium by the BERYLCOAT "D" Process

BERYLCOAT "D" uses a proprietary solution available from Brush Wellman Inc., 17876 St. Clair Ave., Cleveland, OH 44110.

7.2.3 Cleaning

1. Parts that are heavily contaminated with cutting fluids, oil, grease, dirt, etc. should be vapor decreased prior to proceeding with this process. The recommended solvents are Freon TF, Genesolve "D", or 1,1,1-trichloroethane (chlorethane NU).
2. Immerse lightly soiled or vapor degreased parts in clean solvent for 15 minutes with agitation of the parts or the solvent. Ultrasonic cleaning, if available, is preferred. Remove the parts from the solvent and allow to drain for several minutes.
3. Immerse the parts in fresh, clean solvent with agitation for 5 minutes.
4. Dry the parts with a flow of clean dry air or nitrogen. Drying may also be accomplished by placing the parts in an air circulating oven operated at a temperature between 200° and 250°F (90° and 120°C).

   NOTE: Care must be exercised to insure adequate cleaning and drying of recessed areas and holes.
5. Wet parts thoroughly in deionized or distilled water using agitation or flooding. Check for "water-break-free" surfaces. If contamination is still evident as observed by breaks in the water surface when the parts are removed from the water, gently scrub the affected area with a soft, lint-free cloth or tissue saturated with solvent and repeat from Step (2).

7.2.4 Surface Activation Treatment

1. Immerse the parts in a solution consisting of 10 grams of oxalic acid crystals in 100 ml of distilled or deionized water for 20 minutes with periodic agitation. The parts and solution must be at room temperature. The oxalic acid solution should be freshly prepared and should not be stored or held over from one treatment to another. The action of the solution will deteriorate with use and the solution should be discarded after processing approximately 200 square inches of surface area per liter of solution.
2. Wash the parts thoroughly in clean deionized or distilled water using agitation or flooding. Passivate immediately.

7.2.5 Passivation

1. Immerse the parts in a solution consisting of equal parts of BERYLCOAT "D" concentrate and distilled or deionized water for 30 minutes. The BERYLCOAT "D" passivation solution must be thoroughly agitated immediately prior to immersion of the parts. Gently agitate the parts or the solution periodically during the passivation treatment in order to expose the surface to fresh solution and to dislodge any bubbles that may form on the parts.

2. Wash the parts thoroughly in deionized or distilled water using agitation or flooding for 3 minutes.
3. Immerse the parts in at least three separate baths of fresh deionized or distilled water for 10 minutes at a time to remove all traces of BERYLCOAT "D" passivation solution.
4. Dry the parts in a flow of clean dry air or nitrogen. Drying may also be accomplished by placing the parts in an air circulating oven operated at a temperature between 200° and 250°F (90° and 120°C).

7.2.6 General Processing Recommendations

1. Thoroughly agitate the BERYLCOAT "D" concentrate before mixing.
2. Undiluted BERYLCOAT "D" concentrate should not be used for surface treating beryllium parts.
3. Do not use tap water.
4. Parts should not be handled with bare hands while carrying out this process. The use of clean, white, lint-free gloves, clean polyethylene or rubber gloves as well as nonmetallic or plastic-coated cleaning fixtures, racks, containers and tongs is recommended. Avoid the use of uncoated metallic processing equipment especially those made of aluminum.
5. Agitation of the parts or solution is required during all immersion operations.
6. All solutions should be maintained at room temperature. Parts must be at room temperature prior to immersion in any of the working solutions.

7.3 Cadmium (Cadmium Plating)

Cadmium or cadmium plated steel can be made bondable by electroplating with nickel or silver [1,2] followed by a nonchlorinated abrasive scouring. (Refer to the following sections for the appropriate metal.) Cadmium plated steel may also be treated by either the zinc phosphate or chromate conversion coating processes [3].

7.3.1 Preparation of Cadmium and Cadmium Plating by Abrasive Scouring [1]

The parts should be degreased with trichloroethane followed by scouring with a commercial, nonchlorinated abrasive cleaner such as BABO or AJAX. The parts should be rinsed in distilled water and dried with clean, filtered air at room temperature.

7.4 Chromium (Chromium Plate)

Chromium and chromium plate can be made bondable by vapor degreasing with methyl chloroform, trichloroethylene or perchloroethylene, abrasive cleaning or by a hydrochloric acid treatment [2].

7.4.1 Preparation of Chromium and Chromium Plating by Abrasive Cleaning

Solvent clean the surfaces using the above solvents or acetone, mineral spirits or naphtha before and after abrading with 80 to 180 grit paper.

7.4.2 Preparation of Chromium and Chromium Plating by the Hydrochloric Acid Process

Solvent clean the parts as above. Follow by immersion in a 50% hydrochloric acid solution at 200°F for 1 to 5 minutes. Thoroughly rinse with deionized water and dry at 150°F.

7.5 Gold

Gold may be made bondable by either abrasive cleaning or abrasive scouring [1].

7.5.1 Preparation of Gold by Abrasive Cleaning

Solvent clean by vapor degreasing. Avoid immersion and wiping unless these steps can be followed by vapor rinsing or spraying. Then abrade lightly with 180 to 240 grit paper. Solvent clean as above.

7.5.2 Preparation of Gold and Gold Plate by Abrasive Scouring

Scrub the surfaces with distilled water and nonchlorinated scouring powder. Rinse thoroughly with distilled water. Surfaces should be water-break-free.

7.6 Nickel and Nickel Alloys

Nickel parts may be made bondable by abrasive cleaning, nitric acid etching or by a sulfuric-nitric pickling process. Nickel plated parts should not be etched or sanded.

7.6.1 Preparation of Nickel by Abrasive Cleaning [1]

Solvent clean, preferably by vapor degreasing. Then abrade with 180 to 240 grit paper, or grit blast with 40 mesh aluminum oxide abrasive. Solvent clean as above.

7.6.2 Preparation of Nickel by the Nitric Acid Etch Process [1]

Vapor degrease in trichloroethylene vapors. Then etch the parts for 4 to 6 seconds at room temperature (approximately 68°F) in concentrated nitric acid (specific gravity 1.41). Wash well in cold and hot water, followed by a distilled water rinse. Air dry at 104°F (40°C).

7.6.3 Sulfuric-Nitric Acid Pickle [1]

Immerse the parts for 5 to 20 seconds at room temperature in a solution consisting of 30 grams of sodium chloride, 1.5 liters of sulfuric acid (60° Be'), 2.25 liters of nitric acid (40° Be') and 1 liter of water. Follow by rinsing in cold water. Immerse the parts in a 1 to 2% ammonia solution for a few seconds. Rinse thoroughly in distilled water and dry at a temperature of up to 150°F (66°C).

7.7 Nickel Plated Parts [1]

7.7.1 Preparation of Nickel Plated Parts by Abrasive Cleaning

Nickel plated parts may be cleaned by lightly scouring with a nonchlorinated commercial cleaner, rinsing with distilled water, drying at a temperature under 120°F (49°C) and then priming or bonding as soon as possible.

7.7.2 Nickel Alloys

Nickel-based alloys such as Monel®, Inconel®, and Duranickel® reportedly can be treated using methods recommended for treating stainless steel [4]. See Chapter 4.

7.7.3 Inconel

Inconel is a registered trademark of Special Metals Corporation. Inconel refers to a family of austenitic nickel–chromium-based high-performance alloys; these alloys are oxidation and corrosion resistant.

The process is known as sol–gel process and was developed by the Boeing Company as a surface pretreatment for aluminum, titanium, and steel for the repair of aircraft structures at depot and field as well as OEM levels. The process was evaluated by the U.S. Air Force in conjunction with the U.S. Navy, U.S. Army, and the Boeing Company (see references 37 and 38 in Chapter 2). The process is described as a high-performance surface preparation for adhesive bonding of aluminum alloys, steel, titanium, and composites (see reference 39 in Chapter 2).

AC-130 is a sol–gel process that promotes adhesion as a result of the chemical interaction at the interfaces between the adherend and the adhesive or primer. The AC-130 surface prebond surface treatment is a Boeing Company Licensed Product Under Boegel-EPH provided by Advanced Chemistry and Technology, Inc. and registered with the U.S. Patent and Trademark Office.

7.8 Preparation of Inconel Alloys by the AC-130 Sol–Gel Process

7.8.1 Materials

1. AC-130 a four-part system or AC-130-2 a two-part system. AC-Tech 77341 Anaconda Ave, Garden Grove, CA92841
2. Wipers, cheesecloth, gauze, or clean cotton cloth
3. Sanding paper/discs 180-grit, or finer aluminum oxide
4. 3 M Scotch-Brite 2 in (5 cm) or 3 in (7.6 cm) medium grit "Roloc" discs
5. Solvents
6. Bonding primer
7. Proper protective equipment, such as protective gloves, respirators, and eye protection, must be worn during these operations.

7.8.2 Facilities Control

1. Air used in this process shall be treated and filtered so that it is free of moisture, oil, and solid particles.
2. Application of the surface preparation material and primer shall be conducted in an area provided with ventilation.
3. Grinders used shall have a rear exhaust with an attachment to deliver the exhaust away from the part surface.
4. Sanding tools shall have a random orbital movement.

7.8.3 Manufacturing Controls

1. Parts to be processed shall be protected from oil, grease, and fingerprints.
2. Mask dissimilar metals and neighboring regions where appropriate.
3. Apply bond primer within 24 hours of AC-130 application. Cool parts to room temperature.
4. Apply adhesive within 24 hours of AC-130 application.
5. Contain grit and dust residues generated during mechanical deoxidization processes.

7.8.4 Storage

1. Sol–gel kits are considered to be time- and temperature-sensitive and shall be stored in accordance with the supplier's recommendations.

7.8.5 Processing

1. **Precleaning**: Remove all foreign material materials from the surface as needed.
2. Prepare the AC-130 or 130-2 in accordance with the manufacturer's instructions provided in each kit. Scale up for the size of the part as necessary, e.g., 1 liter of solution per 0.9 m^2 (10 ft^2) to be coated. Do not treat the surface with the AC-130 or 130-2 before 30 min induction time is complete. (Induction time is defined as the time period after all the AC-130 components have been mixed, but before the mixed solution is active.) Do not treat surfaces after the 10-hour maximum pot-lifetime has expired.

3. Deoxidation of the surface may be accomplished by grit blasting, sanding, mechanical Scotch-Brite, or manual Scotch-Brite.
   a. **Grit Blasting**: Using alumina grit, grit blast an area slightly larger than the bond area. Use 2–5 atm (30–80 psig) oil-free compressed air or nitrogen. Slightly overlap the blast area with each pass across the surface until a uniform matte appearance has been achieved.
   b. **Sanding**: Using a sander or high-speed grinder connected to oil-free nitrogen or compressed air line, thoroughly abrade the surface with abrasive paper for 1–2 min over a 15 × 15 cm (6 × 6 in) section covering the entire surface uniformly, following the manufacturer's instructions.
   c. **Mechanical Scotch-Brite Abrading**: Replace the abrasive paper as described in Section 7.8.1.4 with a Scotch-Brite abrasive disc and abrade surface as described in Section 7.8.1.4.
   d. **Manual Scotch-Brite**: Thoroughly abrade the surface with a very fine Scotch-Brite pad for a minimum of 1–2 min as described in Section 7.8.1.4. Remove loose grit residue with a clean, dry, natural bristle brush or with clean oil-free compressed air or nitrogen.
4. **Application of AC-130**: Apply AC-130 or 130-2 solution as soon as possible after completion of the deoxidization process. The time between completion of deoxidization and application of AC-130 shall not exceed 30 min. The application of the AC-130 or 130-2 may be accomplished by any of the following methods: spray application, manual application using a clean natural bristle brush or swabbing with a clean wiper, cheesecloth or gauze, or by immersion. Apply solution generously, keeping the surface continuously wet with the solution for a minimum period of one minute. Surface must not be allowed to dry and should be covered with fresh solution at least one time during the solution application period. Allow the coated surface to drain for 5–10 min. If there is any surplus AC-130 solution collected in crevices, pockets, or other contained areas, use filter compressed air to lightly blow off excess solution while leaving a wet film behind. Do not splatter this excess solution onto adjoining surfaces.
5. **Drying**: Allow the coated part to dry under ambient conditions for a minimum of 60 min. Minimize contact with the part during this time, as the coating may be easily damaged until fully cured. Exact drying time will depend on the configuration of the part and room conditions (temperature and humidity).
6. **Bonding**: Apply primer and/or adhesive within 24 hours of AC-130 or 130-2 application. Keep the surface clean during the entire operation.

7.8.6 Acceptable Results

An acceptable AC-130 or 130-2 coating is smooth and continuous without evidence of surface contamination or defects. Dark areas caused by draining of the sol–gel solutions are acceptable.

7.9 Platinum

Platinum may be made bondable by either abrasive cleaning or abrasive scouring [1].

7.9.1 Preparation of Platinum by Abrasive Cleaning

Solvent clean, preferably by vapor degreasing. Avoid immersion and wiping, unless these steps are followed by vapor rinsing or spraying. Then abrade lightly with 180 to 240 grit paper, followed by a solvent cleaning as above.

7.9.2 Preparation of Platinum by Abrasive Scouring

Scrub the surfaces with distilled water and a nonchlorinated scouring powder. Rinse thoroughly with distilled water (surfaces should be water-break-free).

7.10 Silver

Silver can be made bondable by abrasive cleaning or by abrasive scouring as described for platinum [1].

7.11 Uranium

Uranium is a radioactive metal and must be handled with extra care. Local safety regulations should be reviewed with regard to handling and disposal of waste products before working with this metal. The oxide of uranium is dark colored and loosely adhered and must be removed before bonding. Uranium oxidizes very rapidly; a clean surface in contact with air will turn blackish in a matter of minutes. Three methods have been reported to be satisfactory for cleaning uranium [1,5].

7.11.1 Preparation of Uranium by Abrasive Cleaning [5]

This method involves the sanding of all surfaces in a pool of the adhesive to be used for the bonding. This is done to prevent reoxidation of the cleaned surface upon contact with the air. This method has been used with epoxy- and polyurethane-based adhesives. If all of the oxide is not removed during the sanding operation, the uranium will oxidize under the adhesive even though it is not in contact with the air, and the bonds will fail.

7.11.2 Preparation of Uranium by the Acetic Acid–Hydrochloric Acid Process [1]

This method is to be used with aluminum-filled adhesives when no primer or other surface coating is to be applied. The parts are pickled in a bath consisting of 1:1 nitric acid:water. The parts are then rinsed briefly in distilled water and immersed in a bath consisting of 9:1 acetic acid:hydrochloric acid for 3 minutes. Rinse the parts briefly in distilled water, and rinse in acetone. Air dry the parts.

Note that 200 ml of the acetic acid-hydrochloric acid bath can accommodate no more than 4 bars [$4.5 \times 1 \times 0.125$ inches (22.5 in^2)] without a vigorous reaction taking place and a black film forming on the uranium surface.

7.11.3 Preparation of Uranium by the Nitric Acid Process [1]

This method is to be used when a primer or other surface coating is to be applied. The parts are pickled in a bath consisting of 1:1 nitric acid:water. Rinse the parts with acetone and immerse in a bath consisting of 1.0 gram purified stearic acid dissolved in 95 to 99 ml of acetone and 1 to 5 ml of nitric acid. Air dry and then rinse with a carbon tetrachloride spray. Caution: Use with proper safety techniques. Air dry. If parts are to be stored, this should be done in distilled water in a polyethylene container at 140°F (60°C).

7.12 Zinc

Zinc is generally used in the galvanizing of metals and can be bonded to by one of the following processes [1].

7.12.1 Preparation of Zinc by Abrasion Cleaning

Grit or vapor blast the surface or abrade with 100 grit emery cloth. Vapor degrease in trichloroethylene vapors. Dry for at least 2 hours at room temperature, or for 15 minutes at 200°F (93°C) to remove all traces of the trichloroethylene.

7.12.2 Preparation of Zinc by Acid Etch Processes

The parts are vapor degreased in trichloroethylene vapors, abraded with a medium grit emery paper and revapor-degreased. The parts are then etched in a bath consisting of 15 parts by volume concentrated hydrochloric acid or acetic acid and 85 parts by volume distilled water at room temperature for 2 to 4 minutes. The parts are then rinsed with warm tap water, rinsed again with distilled or deionized water and dried in an oven at 150° to 160°F (66° to 71°C).

7.12.3 Preparation of Zinc by the Sulfuric Acid/Dichromate Etch Processes

The parts are degreased in trichloroethylene vapors and then immersed in a bath consisting of 2 pbw concentrated sulfuric acid (specific gravity 1.84), 1 pbw crystalline sodium dichromate, and 8 pbw distilled water at 100°F (38°C) for 3 to 6 minutes. The parts are rinsed in running tap water, again in distilled water and dried in air at 104°F (40°C).

7.13 Rare Metals

CAUTION: THE USE OF SODIUM DICHROMATE IS CONSIDERED A HEALTH HAZARD AND IS CONSIDERED A CARCINOGEN. CHECK WITH THE PROPER GOVERNMENTAL AGENCY BEFORE CONSIDERING THE USE OF THIS METHOD.

The rarer metals such as columbium, copper-beryllium, palladium, rhenium, tantalum and zirconium were found to be bondable when they are vapor degreased in perchloroethylene vapors [6]. These rare earth metals can be prepared by the Openair©Plasma process.

7.14 Metal Matrix Composites

Very little information is available on the surface preparation of the materials known as metal matrix composites. It is of interest that McNamara et al. [7] found that many of the standard processes for treating aluminum will work well with silicone carbide (SIC) reinforced aluminum composite materials.

References

[1] A.H. Landrock, Surface preparation of adherends, Chapter 4 Adhesives Technology Handbook, Noyes Publications, Park Ridge, NJ, 1985.
[2] MIL-HDBK-691B, Military Standardization Handbook, Adhesive Bonding, Chapter 5 (1987).
[3] R.C. Snogren, Miscellaneous metals, Chapter 11 Handbook of Surface Preparation, March 12, Palmerton Publishing Co. Inc., NY, NY, 1974.
[4] R.E. Keith, et al., Adhesive Bonding of Nickel and Nickel-Base Alloys, NASA TMX-53428 October (1965).
[5] R.F. Wegman, M.J. Bodnar, Bonding uranium 238, Adhesives Age 3 (6) (1960) 34–36.
[6] R.F. Wegman, M.J. Bodnar, Bonding rare metals, Machine Design 31 (20) (1959) 139–140.
[7] D. McNamara, A. Desai, T. Fritz, Adhesive bonding of SiC-reinforced aluminum adherends, Proceedings of the Fifth International Joint Military/Government-Industry Symposium on Structural Adhesive Bonding, U.S. Army ARDEC, Picatinny Arsenal, NJ, 3–5 November 1987, pp. 187–205.

8 Plastics

8.1 Introduction

The methods for surface preparation discussed in this chapter are either simple abrasion methods or chemical treatments that have not been developed to the point where production process specifications can be prepared in detail. The methods have been published in the literature but not enough comparative data are available to make recommendations on their selection for use. Potential users should conduct their own investigations of the suitability of a method for a particular application. Chemicals should be checked for safety and health hazards. Except where otherwise referenced, the methods presented in this chapter have been taken from MIL-HDBK 691 B, *Military Standardization Handbook Adhesive Bonding.*

The surface preparation of three classes of plastic materials will be considered, these include:

1. Organic matrix composites,
2. Thermoset materials, and
3. Thermoplastic materials.

CAUTION: THE USE OF CHROMATES ARE CONSIDERED A HEALTH HAZARD AND ARE CONSIDERED CARCINOGENIC. CHECK WITH THE PROPER GOVERNMENTAL AGENCY BEFORE CONSIDERING THE USE OF THESE METHODS.

8.2 Organic Matrix Composites

Composites are defined as systems consisting of two or more constituents, each of which is distinguishable at the microscopic level. These constituents keep their individual identity in the composite and are separated by a detectable interface. One of the constituents acts as a reinforcing agent while the other acts as the matrix or binder. The properties of the composite itself are derived from the combination of the individual properties of the constituents modified by their synergistic effect. The composite materials of interest in this chapter will be limited to organic polymeric matrix materials with "advanced" high performance reinforcements such as fiber glass, graphite and aramid fibers.

Fiber glass is a product produced as a continuous monofilament bundle by drawing molten glass through a multihole bushing. The size of the holes will determine the diameter of the fiber.

Surface Preparation Techniques for Adhesive Bonding.

A common type of graphite fiber is produced by the pyrolysis of polyacrylonitrile fibers under tension at temperatures of 3,200° to 5,000°F (1,760° to 2,760°C) in a controlled atmosphere. The properties of the fibers are a function of both the tension maintained and the temperature at which the pyrolysis is conducted.

Aramid is the generic name for a class of polyamide materials. The principal aramid fiber of concern here is Kevlar. These fibers are produced by conventional textile spinning methods. The fibers are of interest in composite applications because of their outstanding combination of low density, high strength and high stiffness.

There are two major categories of organic matrix materials, thermosets and thermoplastics. To date, thermosets have been the dominating materials for structural applications. Thermoplastic matrix materials are just now coming into popular use. These thermoplastic matrix materials are generally solvent-welded or joined by ultrasonic or spin bonding.

The principal role of the matrix material in a composite is to absorb and transmit loads to the reinforcement fibers. The matrix material also controls many other properties of the composite, such as viscoelastic behavior, creep, stress relaxation, in-plane and inter-laminar shear, as well as chemical, thermal, electrical and environmental aging characteristics. The role of the fibers in the composite is primarily to carry the load through the structure.

There are basically two major techniques for preparation of thermoset matrix composites for bonding. Both of these techniques involve the removal of the original resin surface, leaving a fresh, slightly roughened, uncontaminated faying surface. The removal of surface gloss is generally enough to also remove any surface contamination and to expose a new resin-reinforcement surface. Care must be taken not to damage the fiber. The two techniques are either some form of abrasion or a peel ply.

8.2.1 Peel Ply

Peel ply is an extra layer of fabric material which is laid upon the outer surface of the composite during fabrication. This layer is intended to be peeled off at some future time prior to bonding. The peel ply is a woven fabric, glass, nylon or other synthetic material. During the cure cycle of the fabrication process of the composite, the peel ply fabric absorbs some of the matrix resin and becomes an intergral part of the laminate. When the bonding is to be done, the peel ply is removed, fracturing the resin between it and the first layer of reinforcement. This leaves a fresh, clean, roughened surface of matrix resin to which the adhesive is applied.

The properties of the peel ply fabric will influence the bonding properties of the joint. Matienzo et al. [1] investigated two peel ply materials for their effect on the surface preparation of graphite-epoxy laminates. Peel ply A was a polyester fabric having a thickness of 0.005 inch and a filament diameter of 24.8 μ, while peel ply B was a 0.008 inch thick nylon fabric having a filament diameter of 30.8 μ. Bond strengths to the peel ply A prepared surface averaged 2,372 psi while those to the peel ply B prepared surfaces averaged 2,656 psi. The nylon peel ply B also completely blocked any transfer of silicone mold release agent. Pocius and Wenz [2] also evaluated peel ply materials and reported that abrasion techniques were better than the peel ply.

8.2.2 Abrasion

Composites are often prepared for adhesive bonding by abrasion of the surface to remove the surface gloss and to roughen the surface. This must be done without damaging the reinforcing fibers.

Pocius and Wenz [2] evaluated a number of mechanical methods of preparing graphite-epoxy laminates for bonding. These methods included peel ply materials, machine sanding, grit blasting and some SCOTCH-BRITE methods. Their work showed that abrasion methods were better than peel ply methods. The two best methods in this study involved machine sanding using orbital sanders. The first method used 150 grit silicon carbide FRE-CUT coated abrasive paper mounted on an orbital pad sander having a 3/16 inch diameter orbit, using a 4 × 9 inch pad area. The other method used a SCOTCH-BRITE BSD DISC. This is described as a high loft (1 inch thick) 5 inch diameter "Brown Stripper" mounted on a random orbital disc sander operated at 10,000 rpm.

Wade [3] developed repair procedures for graphite composite skin panels on the spaceship *Columbia*. This technique first required precleaning to remove foreign materials. This involved hand sanding and solvent wiping. The faying surfaces were then grit blasted to a lusterless appearance with 220 grit aluminum oxide under 25 psig air pressure. The blasted surfaces were then vacuumed and dry wiped clean prior to bonding.

For graphite-polyimide composites, the most popular techniques for preparing the surfaces for bonding appear to be light abrasion followed by solvent cleaning and drying. Progar [4] used the following technique: light grit blasting with 120 grit aluminum oxide followed by an alcohol wash, drying in a forced air oven at 393°F (199°C) for 16 hours and then priming. Deaton and Masso [5] also recommended a light grit blast followed by a methyl ethyl ketone (MEK) wipe and then drying. Steger [6] recommends lightly sanding with a 280 grit sand paper, followed by a methyl ethyl ketone (MEK) wipe and priming.

Glass reinforced composites generally are sanded lightly with 80 to 120 grit abrasive paper to roughen the surface, break the surface gloss and to remove release agents. Sanding should stop before the glass fibers are damaged. Grit blasting with 120 grit aluminum oxide may also be employed. After abrasive cleaning the surface may be vacuumed or wiped with a solvent such as MEK, acetone, toluene, trichloroethylene, trichloroethane or Freon TF depending upon the type of organic matrix.

Aramid fiber composites can also be prepared for bonding by the peel ply or the abrasive techniques. Abrasion of Kevlar laminates can be accomplished by using aluminum oxide or silicon carbide abrasive paper, 80 to 100 grit, at a surface speed of 4,000 to 5,000 ft/min using wet paper. Care should be taken not to damage the fiber.

8.3 Thermoset Materials

Thermoset materials are defined as materials that will undergo or have undergone a chemical reaction by the action of heat, catalyst, ultraviolet light, etc., resulting in a relatively infusable state [7]. Once the material has undergone the chemical reaction it cannot repeatedly be softened upon reheating and hardened upon cooling. These materials are not soluble in the normal cleaning solvents such as acetone, methyl ethyl ketone, toluene, trichloroethylene, perchloroethylene, alcohols, etc. These solvents are generally used to remove or reduce the amount of release agents which are often found on the surface of the parts made of thermoset materials.

The basic cleaning process used for the majority of thermoset materials involves cleaning, with one of the solvents mentioned above, followed by an abrasive treatment using a fine abrasive paper, abrasive blast or metal wool followed by another solvent cleaning. Variations of this basic cleaning technique have been used on the following thermoset materials: epoxies, phenolics, polyesters, diallyl-phthalates, melamines, polyimides, urea-formaldehydes, and some polyurethanes and silicones [8].

8.4 Thermoplastic Materials

Thermoplastic materials can be reworked by heating and cooling provided that the temperature is not high enough to cause decomposition. These materials generally require special treatments to chemically or physically modify their surface to make them bondable. Some adhesive or primers may be required for specific polymers. The following treatment methods are taken from MIL-HDBK 691 B unless otherwise referenced.

8.5 Acetal Copolymer (Celcon)

8.5.1 Preparation of Acetal Copolymer by the Potassium Dichromate/Sulfuric Acid Etch Process

1. Wipe parts with acetone.
2. Air dry.
3. Etch surface for 10 to 15 seconds in a solution containing

Sulfuric acid	400 pbw
Potassium dichromate	11 pbw
Water, deionized	44 pbw

4. Immediately flush parts with tap water.
5. Rinse with deionized water.
6. Oven dry at 140°F (60°C).

8.5.2 Preparation of Acetal Copolymer by the Hydrochloric Acid Etch (Nonchromate) [9] Process

1. Immerse parts for five minutes in concentrated hydrochloric acid at room temperature.
2. Use a glass rod to move fresh acid into contact with the acetal copolymer. About 8–13 μm of acetal copolymer is removed per minute of the etching process.
3. Rinse and dry at room temperature for 4 hours.

8.6 Acetal Homopolymer (Delrin)

8.6.1 Preparation of Acetal Copolymer by the Potassium Dichromate/ Sulfuric Acid Process

1. Wipe with acetone or MEK.
2. Etch surface for 10 to 20 seconds at room temperature (68° to 86°F, 20° to 30°C) in a solution containing

Potassium dichromate	75 g or 15 pbw
Sulfuric acid, concentrated	1,500 g or 300 pbw
Tap water	120 g or 24 pbw

   Dissolve the potassium dichromate in the tap water, then add the sulfuric acid in increments of about 200 g, stirring after each addition.
3. Rinse in running tap water for at least 3 minutes.
4. Rinse in distilled or deionized water.
5. Dry in an air circulating oven at 100°F (38°C) for about 1 hour.

8.6.2 Preparation of Acetal Copolymer by the DuPont® Three-Step Method [9]

Apparatus

Tanks

1. Tanks used for Step 1 solution should be constructed of PVC, polypropylene (PP) high density polyethylene (HDPE), or glass.
2. Tanks used for Step 2 solution should be constructed of stainless steel, mild steel, or glass.
3. Tanks used for Step 3 solution should be constructed of PVC, PP, HDPE, or glass.

Heating Elements

1. Heating elements used for Step 1 tanks should be either quartz heaters or elements coated with Teflon®.
2. Heating elements used for Step 2 tanks should be either stainless steel or titanium coated with Teflon® or quartz.

Materials

Treating Solutions

1. Step 1 Solution—71% orthophosphoric acid (H_3PO_4)
   Mix 5 pbv of commercial (85%) H_3PO_4, and pbv deionized water.
2. Step 2 Solution—12% sodium hydroxide (NaOH)
   Dissolve 120 g NaOH in 1 liter of deionized water.
3. Step 3 Solution—1% acetic acid
   Add 20 ml of glacial acetic acid to 1 liter of deionized water.

Processing

Degrease

If the surface of the part is contaminated (mold release, finger prints, traces of oil, etc.), then the part should be degreased prior to the dipping.

Treatment

1. **Step 1**: Immerse the part into Solution 1 (Section 2.1.1) at 70°C for five minutes. Rinse the part thoroughly with 23°C water for one minute.
2. **Step 2**: Immerse the part into Solution 2 (Section 2.1.2) at 100°C for six minutes. Rinse the part thoroughly with 23°C water for one minute.
3. **Step 3**: Immerse the part into Solution 3 (Section 2.1.3) at 80°C for 10 min. Rinse the part thoroughly with 23°C water for one minute followed by a supplementary rinse in deionized water.

8.7 Acrylonitrile-Butadiene-Styrene

8.7.1 Preparation of ABS by an Abrasive Process

1. Sand with a medium grit sandpaper.
2. Wipe free of dust.
3. Dry in an oven at 160°F (71°C) for 2 hours.
4. Prime with Dow Corning A-4094 or General Electric SS-4101.

8.7.2 Preparation of ABS by the Sulfuric Acid/Potassium Dichromate Process

1. Degrease in acetone.
2. Etch for 20 minutes at 140°F (60°C) in a solution containing:

Sulfuric acid, concentrated	26 pbw
Potassium dichromate	3 pbw
Water	11 pbw

3. Rinse in tap water.
4. Rinse in distilled water.
5. Dry in warm air.

8.8 Cellulosics

This group includes cellulose acetate, cellulose acetate butyrate, cellulose nitrate, cellulose propionate and ethyl cellulose. The cellulosics are usually solvent cemented and require no special surface preparation. When a cellulosic is to be bonded to a dissimilar adherend the following surface preparation may be helpful.

1. Solvent-degrease in methyl or isopropyl alcohol.
2. Grit- or vapor-blast, or use 220 grit emery cloth.
3. Solvent-degrease as in Step (1).
4. Heat 1 hour at 200°F (93°C) and apply adhesive while still hot.

8.9 Ethylene-Vinyl Acetate

1. Degrease in methanol.
2. Prime with epoxy adhesive.
3. Fuse into the surface by heating for 30 minutes at 212°F (100°C).

8.10 Fluorinated Ethylene-Propylene (Teflon®, FEP)

8.10.1 Preparation of FEP by the Sodium Naphthalene Complex Process (ASTM D2093)

1. Wipe with acetone.
2. Treat with sodium-naphthalene complex solution for 15 minutes at room temperature. (These solutions are commercially available.) Sodium-naphthalene complex solutions should be kept in tightly-stoppered glass containers to exclude air and moisture. Extreme care should be taken in handling the solutions since they are hazardous, due to the explosive property of the sodium. Directions are given in the literature for making up the solutions which contain sodium metal, naphthalene and tetrahydrofuran in varying proportions.
3. Remove from the solution with metal tongs.
4. Wash with acetone to remove excess organic material.
5. Wash with distilled or deionized water to remove the last traces of metallic salts from the treated surfaces.
6. Dry in an air circulating oven at 99.5°F (37.3°C) for about 1 hour.

8.11 Nylon

8.11.1 Preparation of Nylon by the Abrasive Treatment Process

1. Wash with acetone.
2. Dry.

3. Hand-sand with 120 grit abrasive cloth until gloss is removed.
4. Remove sanding dust with a short-haired stiff brush.

Other solvents which may be used to clean nylon include: methyl ethyl ketone, perchloroethylene, 1,1,1-trichloroethylene, M-17, and Freon TMC.

When bonding nylon to metal an epoxy type adhesive may be used. The nylon part should be cleared as above and then primed. Some primers that have been recommended include: nitrile-phenolic, resorcinol-formaldehyde, vinyl-phenolic and silanes [8].

8.11.2 Preparation of Nylon by the Sulfuric Acid/ Potassium Dichromate Process

1. Solvent-clean with isopropanol or aqueous solution of commercial detergent.
2. Etch for 1 minute at 176°F (80°C) in a solution consisting of:

Sulfuric acid, concentrated	375.0 g
Potassium dichromate	18.5 g
Water	30.0 g

3. Rinse in distilled water.
4. Dry.

8.11.3 Preparation of Nylon by the Alternate Abrasive Treatment Process

1. Solvent-clean with isopropanol or aqueous solution of commercial detergent.
2. Sand.
3. Acetone wipe.
4. Dry.

8.12 Phenyl Oxide-Based Resins (Noryl® Polyaryl Ethers)

8.12.1 New Method

Preparation of Phenylene Oxide-Based Resins by Abrasion [10]

1. Clean with isopropanol or aqueous solution of most commercial-based detergents.
2. Sand or vacuum blast.
3. Rinse in distilled water.
4. Dry.

8.12.2 New Method

Preparation of Phenylene Oxide-Based Resins by the Acid Etch Procedure [10]

1. Solvent-clean with isopropanol or aqueous solution of most commercial detergents.

2. Etch with a solution consisting of 375 g of sulfuric acid, concentrated, 18.5 g of potassium dichromate, and 30 g distilled water. If the chromate acid etch is used, etch one minute at 80°C in the solution.
3. Rinse in distilled water.
4. Dry.

8.13 Polyaryl Sulfone (Astrel®)

8.13.1 Preparation of Polyaryl Sulfone by Solvent Washing

1. Triple wash (3 successive washes) in a 65/35 mixture by volume of Freon PCA and reagent grade isopropyl alcohol. This mixture is also available from Du Pont as Freon TP-35, except that Freon TF solvent is used instead of Freon PCA.
2. Dry at 150°F (66°C) in an air circulating oven.

8.13.2 Preparation of Polyaryl Sulfone by Abrasive Treatment

1. Ultrasonically clean with an alkaline etching solution.
2. Rinse in cold water.
3. Alcohol wash.
4. Sandblast with 150-mesh silica sand.
5. Alcohol wash.
6. Dry with nitrogen.

8.13.3 Preparation of Polyaryl Sulfone by the Sulfuric Acid/ Potassium Dichromate Process

1. Ultrasonically clean in an alkaline cleaning solution.
2. Rinse in cold water.
3. Immerse for 15 minutes at 150° to 160°F (66 to 71°C) in a solution consisting of:
   Sodium dichromate 3.4% by wt
   Sulfuric acid 96.6% by wt
4. Wash in cold water.
5. Dry at 150°F (66°C) in an air-circulating oven.

8.14 Polycarbonate (Lexan®, Calibre®, and Tuffak®)

8.14.1 Preparation of Polycarbonate by Flame Treatment

1. Wipe with ethanol to remove dirt and grease.
2. Pass the part through the oxidizing portions of a propane flame. Treatment is complete when both sides are polished to a high gloss, free of scratches and visible flaws. The process usually requires 5 to 6 passes on both sides.
3. Allow the part to cool for 5 to 10 minutes before bonding.

8.14.2 Preparation of Polycarbonate by Hot Air Treatment

1. Place the part in an air-circulating oven at 160°F (71°C) for 1 hour.
2. Five to 10 minutes after removal from oven, parts are cool enough to bond.

8.14.3 Preparation of Polycarbonate by Abrasive Treatment

1. Wipe with ethanol, methanol, isopropanol, petroleum ether, or heptane. CAUTION: polycarbonates are very prone to crazing or cracking, particularly when exposed to solvents while under stress. Extreme care is necessary when using solvents to clean polycarbonates.
2. Air dry.
3. Sand with fine (120 to 400 grit maximum) abrasive cloth or sandpaper.
4. Remove sanding dust with a clean dry cloth or a stiff, short-hair brush.
5. Repeat solvent wipe.

8.15 Polyethylene

8.15.1 Preparation of Polyethylene by the Potassium Dichromate/ Sulfuric Acid Process

1. Wipe with acetone, MEK or xylene.
2. Immerse for 60 to 90 minutes at room temperature, or 30 to 60 seconds at 160°F (71°C) in a solution consisting of:

Potassium dichromate	75 g	15 pbw
Tap water	120 g	24 pbw
Sulfuric acid, concentrated	1,500 g	300 pbw

   Dissolve the potassium dichromate in the tap water, then add the sulfuric acid slowly in increments of about 200 g, stirring after each addition.
3. Rinse in running tap water for at least 3 minutes.
4. Rinse in distilled or deionized water.
5. Dry in an air circulating oven at 100°F (38°C) for about 1 hour.

8.15.2 Preparation of Polyethylene by the Flame Treatment Method

1. Pass the oxidizing flame of the oxyacetylene burner over the faying surface until it appears glossy.
2. Scour lightly with soap and water.
3. Wash with distilled water.
4. Dry at room temperature.

This method requires very careful control to prevent heat warping.

8.15.3 Introduction

Rosty et al. [9,10] introduced some method for treating low density polyethylene. These include the following.

8.15.4 Newer Methods of Treating Low Density Polyethylene

Rosty et al. [9,10] introduced some newer methods for treating low density polyethylene. These include the following.

8.15.5 Preparation of Low Density Polyethylene by the Potassium Dichromate/Sulfuric Acid Process

See above, except that the immersion time at room temperature was 1 week.

1. Wipe with acetone, MEK, or xylene.
2. Immerse the part for one week at room temperature in a solution consisting of:
   Potassium dichromate 75 g (15 pbw)
   Tap water 120 g (24 pbw)
   Sulfuric acid, concentrated 1,500 g (300 pbw)
   Dissolve the potassium dichromate in the tap water, then add the sulfuric acid slowly in increments of about 200 g, stirring after each addition.
3. Rinse in running water for at least three minutes.
4. Rinse in distilled or deionized water.
5. Dry in an air-circulating oven at 38°C (100°F) for about one hour.

8.15.6 Preparation of Low Density Polyethylene by the Bleach-Detergent Process

1. Immerse the part for 1 week at room temperature in a solution consisting of:
   Bleach 73.5 ml
   Detergent 3.5 g
2. Rinse in tap water.
3. Dry at room temperature with dry air.

8.15.7 Preparation of Low Density Polyethylene by the Lead Dioxide–Sulfuric Acid Process

1. Immerse the part for 4 hours at 160°F (71°C) in a solution consisting of:
   Lead dioxide 6.3 g
   Water 12.0 g
   Sulfuric acid, concentrated 82.0 ml
2. Rinse in tap water.
3. Dry at room temperature with dry air.

8.15.8 Preparation of Low Density Polyethylene by the Potassium Iodate–Sulfuric Acid Process

1. Immerse the part for 1 week at room temperature in a solution consisting of:
   Potassium iodate 5.5 g
   Tap water 12.0 g
   Sulfuric acid, concentrated 82.0 ml

2. Rinse in tap water.
3. Dry at room temperature with dry air.

8.16 Polymethylmethacrylate (Plexiglas or Lucite)

8.16.1 Preparation of Polymethylmethacrylate by Abrasive Treatment

1. Wipe with methanol, acetone, isopropanol, or use a detergent.
2. Abrade with fine grit (180 to 400 grit) sandpaper or emery paper, or use abrasive scouring with a small amount of water, dry grit blasting or wet abrasive blasting.
3. Wipe with a clean, dry cloth to remove particles.
4. Repeat solvent wipe.

8.17 Polymethylpentane

8.17.1 Preparation of Polymethylpentane by Abrasive Treatment

1. Solvent clean in acetone.
2. Grit or vapor blast, or use 100 grit emery cloth.
3. Solvent clean in acetone.

8.17.2 Preparation of Polymethylpentane by the Potassium Dichromate/Sulfuric Acid Process

1. Solvent clean in acetone.
2. Immerse for 1 hour at 140°F (60°C) in a solution consisting of:

Potassium dichromate	3 pbw
Water	11 pbw
Sulfuric acid, concentrated	26 pbw

3. Rinse in tap water.
4. Rinse in distilled water.
5. Dry in warm air.

8.17.3 Preparation of Polymethylpentane by the Potassium Permanganate Process

1. Solvent clean in acetone.
2. Immerse for 5 to 10 minutes at 194°F (90°C) in a saturated solution of potassium permanganate acidified with sulfuric acid.
3. Rinse in tap water.
4. Rinse in distilled water.
5. Dry in warm air.

8.18 Polyphenylene Sulfide (Ryton®)

8.18.1 Preparation of Polyphenylene Sulfide by Abrasive Treatment

1. Solvent degrease in acetone.
2. Sandblast.
3. Solvent degrease in acetone.

or

1. Wipe with ethanol soaked lintless paper.
2. Sand with 120 grit sandpaper.
3. Clean off dust with a stiff bristle brush.

8.19 Polypropylene [11,12]

8.19.1 Preparation of Polypropylene by the Potassium Dichromate/ Sulfuric Acid Process

1. Wipe with acetone, MEK, or xylene.
2. Immerse for 1 to 2 minutes at 160°F (71°C) in a solution consisting of:

Potassium dichromate	75 g	15 pbw
Tap water	120 g	24 pbw
Sulfuric acid, concentrated	1,500 g	300 pbw

   Dissolve the potassium dichromate in the water, then add the sulfuric acid in increments of about 200 g, stirring after each addition.
3. Rinse in running tap water for at least 3 minutes.
4. Rinse in distilled or deionized water.
5. Dry in an air circulating oven at 100°F (38°C) for about 1 hour.

8.19.2 Preparation of Polypropylene by the Flame Treatment Method

1. Pass the oxidizing flame of an oxyacetylene burner over the faying surface until it appears glossy.
2. Scour lightly with soap and water.
3. Wash with distilled water.
4. Dry at room temperature.

This method requires very careful control to prevent heat warping.

8.19.3 New Methods for Treating Polypropylene

Preparation of Polypropylene by the Room Temperature Potassium Dichromate/Sulfuric Acid Process

See above, except that the immersion time was 1 hour at room temperature.

1. Wipe with acetone or xylene.
2. Immerse the part for one week at room temperature in a solution consisting of:

Potassium dichromate	75 g (15 pbw)
Tap water	120 g (24 pbw)
Sulfuric acid, concentrated	1,500 g (300 pbw)

   Dissolve the potassium dichromate in the tap water, then add the sulfuric acid slowly in increments of about 200 g, stirring after each addition.
3. Rinse in running water for at least three minutes.
4. Rinse in distilled or deionized water.
5. Dry in an air-circulating oven at 38°C (100°F) for about one hour.

8.19.4 Preparation of Polypropylene by the Bleach-Detergent Process

1. Immerse the part for 1 week at room temperature in a solution consisting of:

Bleach	73.5 ml
Detergent	3.5 g

2. Rinse in tap water.
3. Dry at room temperature with dry air.

8.19.5 Preparation of Polypropylene by the Lead Dioxide–Sulfuric Acid Process

1. Immerse the part for 1 week at room temperature in a solution consisting of:

Lead dioxide	6 g
Tap water	12 g
Sulfuric acid, concentrated	82 ml

2. Rinse in tap water.
3. Dry at room temperature with dry air.

8.20 Polystyrene

8.20.1 Preparation of Polystyrene by Abrasive Treatment

1. Degrease with methyl or isopropyl alcohol.
2. Abrade with 200 grit sand paper.
3. Remove dust particles.

8.20.2 Preparation of Polystyrene by the Sulfuric Acid/Sodium Dichromate Process

1. Degrease with isopropyl or methyl alcohol.
2. Immerse for 3 to 4 minutes at 210° to 220°F (99° to 104°C) in a solution consisting of:

Sulfuric acid, concentrated	90 pbw
Sodium dichromate	10 pbw

3. Rinse thoroughly with distilled water.
4. Dry below 120°F (49°C).

8.21 Polysulfone

8.21.1 Preparation of Polysulfone by the Sulfuric Acid/ Sodium Dichromate Process

1. Ultrasonically clean in alkaline etching solution.
2. Rinse in cold water.
3. Immerse for 15 minutes at 150° to 160°F (66° to 71°C) in a solution consisting of:
   Sodium dichromate 3.4% by wt
   Sulfuric acid, concentrated 96.6% by wt
4. Wash in cold water.
5. Dry at 150°F (66°C) in an air circulating oven.

8.21.2 Preparation of Polysulfone by the Solvent Wash Process

1. Triple wash (3 successive washes) in a 65/35 mixture by volume of Freon PCA and reagent grade isopropyl alcohol.
2. Dry at 150°F (66°C) in an air circulating oven.

8.21.3 Preparation of Polysulfone by Abrasive Treatment [13]

1. Abrade with 80 to 120 grit abrasive.
2. Wipe with Freon 12.

References

[1] L.J. Matienzo, J.D. Venables, J.D. Fudge, J.J. Velton, Surface Preparation of Bonding Advanced Composites, Part 1; Effect of Peel Ply Materials and Mold Release Agents on Bond Strengths, 30th National SAMPE Symposium, 19–21 March 1985, pp. 302–314.

[2] A.V. Pocius, R.P. Wenz, Mechanical Surface Preparation of Graphite-Epoxy Composites for Adhesive Bonding, 30th National SAMPE Symposium, 19–21 March 1985, pp. 1073–1081. Also SAMPE J. 21(5)(1985) pp. 50–58.

[3] H.A. Wade, Field Repair of Graphite Epoxy Skin Panels on the Spaceship Columbia, 29th National SAMPE Symposium, 12–14 April 1983, pp. 249–257.

[4] D.J. Progar, High Temperature Adhesives for Bonding Composites, 11th National SAMPE Technical Conference, 13–15 November 1979, pp. 233–251.

[5] J.W. Deaton, N.A. Masso, Preliminary Evaluation of Large Area Processes for Repair of Graphite/Polyimide Composites, 28th National SAMPE Symposium, 12–14 April 1983, pp. 1425–1442.

[6] V.Y. Steger, Structural Adhesive Bonding Using Polyimide Resins, 12th National SAMPE Technical Conference, 7–9 October 1980, pp. 1054–1059.

[7] ASTM D907 Standard definitions of terms relating to adhesives, Annual Book of ASTM Standards, vol. 15.06, ASTM, Philadelphia, PA.

[8] A.H. Landrock, Surface preparation of adherends, Chapter 4 Adhesives Technology Handbook, Noyes Publications, Park Ridge, NJ, 1985.

[9] Delrin® Design Information Module II No. L10464.

[10] R. Rosty, D. Martinelli, A. Devine, M.J. Bodnar, J. Beetle, Surface Preparation of Polyolefins Prior to Adhesive Bonding, 32nd International SAMPE Symposium, 6–9 April 1987, pp. 456–465, and SAMPE J. 23(4)(1987) pp. 34–37.
[11] R. Rosty, D. Martinelli, A. Devine, M.J. Bodnar, J. Beetle, Preparation of Polyolefin Surfaces for Adhesive Bonding, Technical Report ARAED-TR-86005, U.S. Army Armament Research and Development Center, May 1986.
[12] G.C. Liskay, Development of Manufacturing Methods for Joining Reinforced Thermoplastics (Tomahawk Thermoplastic Composite Wing), 13th National SAMPE Technical Conference, 13–15 October 1981, pp. 592–602.
[13] Sina Ebneajjad, Surface Treatment of Materials for Adhesion Bonding, William Andrew Publisher, Norwich, NY, USA, 2006.

9 Rubbers

9.1 Introduction

The methods presented in this chapter have been taken from MIL-HDBK 691 B, *Military Standardization Handbook "Adhesive Bonding."* These methods have not been adopted for large scale production use and therefore have not been put into a process specification format. Very little data evaluating the methods are available, therefore potential users are cautioned to conduct their own evaluations before adopting a method for use.

Almost all commercially available types of elastomers can be strongly bonded to a large variety of substrates by vulcanization. The preparation of strong reliable bonds to vulcanized elastomers is the topic of this chapter. These strong reliable bonds can be achieved by the use of proper surface preparation.

Vulcanized rubber parts are often contaminated with mold-release agents, talc, dirt, grease, plasticizers, extender oils, or other compounding ingredients that migrate to the surface and interfere with adhesive bonding. Solvent washing and abrading are common surface treatments for most elastomers, but chemical treatment is often required for maximum strength and/or other properties such as aging and weathering. Many synthetic and natural rubbers require "cyclizing" with concentrated sulfuric acid until hairline cracks appear on the surface. Other rubbers require the use of primers for optimum bonding.

Mechanical abrasion is generally accomplished by sanding or buffing the surface with 80 to 240 grit sand paper or a buffing wheel. The dust from the abrasion process is usually removed by wiping with a clean cloth dipped in a suitable solvent. The solvent must be reasonably compatible with the type of rubber being cleaned. If the solvent is very strongly incompatible with the rubber involved, or if too much is used, the rubber will swell excessively, may curl unacceptably, or may be degraded. Particular care must then be taken not to trap solvent in a system that is totally closed, or else the rubber may be damaged.

On the other hand, a mild wipe with a somewhat aggressive solvent may help to tackify the rubber surface. Methyl ethyl ketone (MEK) and toluene are solvents commonly used for cleaning elastomers. MEK is a strong solvent for fluorocarbon elastomers, and excessive curling has been experienced when MEK is used in any quantity with this type of elastomer. Toluene is a more compatible solvent and may be used instead of the MEK. The chlorinated solvents, such as perchloroethane and 1,1,1-trichloroethane, may also be used.

Surface Preparation Techniques for Adhesive Bonding.

Chlorination is another commonly used surface preparation especially with the neoprene rubbers.

9.2 Neoprene

Neoprene is the generic name for polymers of chloroprene (2-chloro-1,3-butadiene). Neoprene rubbers are available in both solid and latex forms.

9.2.1 Preparation of Neoprene by Abrasive Treatment

1. Scrape the surface with a sharp blade to remove gross layers of wax, sulfur and other compounding ingredients which may have floated to the surface during processing.
2. Solvent-wipe with ethyl, isopropyl, or methyl alcohol, MEK, or toluene.
3. Uniformly abrade the surface with 80 to 120 grit abrasive paper. Machine sanding with a "jitterbug" oscillating sander is preferred over hand sanding because the machine sanding produces a more uniform surface with less effort. A buffing wheel may also be used.
4. Solvent wipe again as in Step (2) to remove loose particles.

9.2.2 Preparation of Neoprene by the Cyclization Process

This process is often applied preparatory to bonding with flexibilized epoxy adhesives.

1. Scrape the surface with a sharp blade to remove gross layers of wax, sulfur and other compounding ingredients which may have floated to the surface during processing.
2. Solvent-wipe with ethyl, isopropyl, or methyl alcohol, MEK or toluene.
3. Immerse the rubber surface in concentrated sulfuric acid (specific gravity 1.84) for 5 to 45 minutes or for a period of time determined for each rubber composition which will yield optimum bond strength.
4. Rinse thoroughly with tap water. Hot water is preferable.
5. Rinse thoroughly with distilled water.
6. Neutralize by immersing for 5 to 10 minutes in a 10 to 20% solution of ammonium hydroxide (sodium hydroxide may also be used).
7. Rinse thoroughly with tap water.
8. Rinse thoroughly with distilled water.
9. Dry at temperatures up to 150°F (66°C).
10. Flex the resultant brittle surface of the rubber using clean rubber or plastic gloves so that a finely cracked appearance is produced. Cyclizing hardens and slightly oxidizes the surface, permitting the necessary wetting of the surface by the adhesive. The finely cracked surface indicates that the rubber is ready for bonding. Light lacy lines on the surface indicate insufficient immersion time. Deep, coarse cracks and a thick crusty surface indicate excessive immersion. If immersion is not feasible, the acid may be made up into a thick paste by the addition of barium sulfate (barytes) or CAB-O-SIL (G.L. Cabot Corp.) or 100 pbw concentrated sulfuric acid:5 pbw SANTOCEL C (Monsanto Co.) or its equivalent may be used. A stainless steel or other acid-resistant spatula should be used to apply the paste. After the paste is applied, treatment Steps (4) through (9) should be carried out.

9.2.3 Cyclization (Modified Process [1])

1. Clean the neoprene in toluene.
2. Force dry for one hour at 60°C.
3. Immerse for two minutes in concentrated sulfuric acid.
4. Rinse in tap water.
5. Force dry for one hour at 60°C.

9.2.4 Preparation of Neoprene by the Chlorination Process

1. Scrub the neoprene in 120° to 140°F (49° to 60°C) nonionic detergent solution (2 to 3% by weight).
2. Rinse thoroughly in tap water.
3. Rinse thoroughly in distilled water.
4. Air dry.
5. Immerse for 1½ to 3 minutes in the following solution at room temperature. The solution should be prepared just prior to use, adding the ingredients in the order listed:
   Distilled water 97.0 pbw
   Sodium hypochlorite (as Clorox or Purex) 3.0 pbw
   Hydrochloric acid (specific gravity 1.20) 0.3 pbw
6. Rinse thoroughly with distilled water and dry at a temperature up to 150°F (66°C).

9.3 Ethylene-Propylene-Diene Terpolymer (EPDM)

This rubber, along with the copolymer of ethylene or propylene (EPM), forms a class of rubbers called ethylene-propylene. However, since EPDM, the terpolymer, is in much greater use than EPM, the former is frequently called ethylene-propylene rubber. Its outstanding property is its very high resistance to ozone and weathering. Being less polar, EPDM is relatively difficult to bond.

9.3.1 Preparation of EDPM by Abrasive Treatment

1. Scrape the surface with a sharp blade to remove gross layers of wax, sulfur and other compounding ingredients which may have floated to the surface during processing.
2. Solvent wipe with acelone or MEK.
3. Uniformly abrade the surface with 80 to 120 grit abrasive paper. Machine sanding with a "jitterbug" oscillating sander is preferred over hand sanding because machine sanding produces a more uniform surface with less effort. A buffing wheel may also be used.
4. Solvent wipe again as in Step (2) to remove loose particles.

9.3.2 Preparation of EDPM by the Cyclization Process

This process is often applied preparatory to bonding with flexibilized epoxy adhesives.

1. Scrape the surface with a sharp blade to remove gross layers of wax, sulfur and other compounding ingredients which may have floated to the surface during processing.
2. Solvent wipe with ethyl, isopropyl, or methyl alcohol, MEK or toluene.

3. Immerse the rubber surface in concentrated sulfuric acid (specific gravity 1.84) for 5 to 45 minutes or for a period of time determined for each rubber composition which will yield optimum bond strength.
4. Rinse thoroughly with tap water. Hot water is preferable.
5. Rinse thoroughly with distilled water.
6. Neutralize by immersing for 5 to 10 minutes in a 10 to 20% solution of ammonium hydroxide (sodium hydroxide may also be used).
7. Rinse thoroughly with tap water.
8. Rinse thoroughly with distilled water.
9. Dry at temperatures up to 150°F (66°C).
10. Flex the resultant brittle surface of the rubber using clean rubber or plastic gloves so that a finely cracked appearance is produced. Cyclizing hardens and slightly oxidizes the surface, permitting the necessary wetting of the surface by the adhesive. The finely cracked surface indicates that the rubber is ready for bonding. Light lacy lines on the surface indicate insufficient immersion time. Deep, coarse cracks and a thick crusty surface indicate excessive immersion. If immersion is not feasible, the acid may be made up into a thick paste by the addition of barium sulfate (barytes) or CAB-O-SIL (G.L. Cabot Corp.) or 100 pbw concentrated sulfuric acid:5 pbw SANTOCEL C (Monsanto Co.) or its equivalent may be used. A stainless steel or other acid-resistant spatula should be used to apply the paste. After the paste is applied, treatment Steps (4) through (9) should be carried out.

9.4 Silicone Rubber

These rubbers (polydimethylsiloxane) are completely synthetic materials and find wide use for many applications. They consist of two types—heat vulcanizing and room temperature vulcanizing (RTV). Both types have unique properties unobtainable with organic rubbers, particularly where superior endurance and extended life are required. Silicone rubbers maintain their usefulness from −150° to 500°F (−100° to 316°C). Resistance to oxidation, oils and chemicals is high and their stability against weathering is very good.

9.4.1 Preparation of Silicone Rubber by Solvent Cleaning

1. Sand with a medium-grit sandpaper.
2. Solvent-wipe surfaces with acetone, MEK, ethyl, methyl or isopropyl alcohol, or toluene.

9.4.2 Preparation of Silicone Rubber by Soap-and-Water Treatment

1. Wash with a mild (Ivory) soap.
2. Rinse thoroughly in tap water.

9.4.3 Priming

Primers have been developed for use with the silicone rubbers and should be used in accordance with the manufacturer's instructions.

9.5 Butyl Rubber

Butyl rubber is a copolymer of isobutylene and isoprene. It is a well established specialty rubber with a wide range of applications. Its special properties are: low gas permeability, thermal stability, ozone and weathering resistance and vibration damping. It has a higher coefficient of friction and chemical and moisture resistance than others. Being less polar, butyl rubber is relatively difficult to bond.

9.5.1 Preparation of Butyl Rubber by Abrasive Treatment

1. Scrape the surface with a sharp blade to remove gross layers of wax, sulfur and other compounding ingredients which may have floated to the surface during processing.
2. Solvent wipe with toluene.
3. Uniformly abrade the surface with 80 to 120 grit abrasive paper. Machine sanding with a "jitterbug" oscillating sander is preferred over hand sanding because machine sanding produces a more uniform surface with less effort. A buffing wheel may also be used.
4. Solvent wipe again with toluene to remove loose particles.

9.5.2 Preparation of Butyl Rubber by the Cyclization Process

1. Scrape the surface with a sharp blade to remove gross layers of wax, sulfur and other compounding ingredients which may have floated to the surface during processing.
2. Solvent wipe with toluene.
3. Immerse the rubber surface in concentrated sulfuric acid (specific gravity 1.84) for 5 to 45 minutes [see Step (10)].
4. Rinse thoroughly with tap water. Hot tap water is preferable.
5. Rinse thoroughly with distilled water.
6. Neutralize by immersion for 5 to 10 minutes in a 10 to 20% solution of ammonium hydroxide (sodium hydroxide may also be used).
7. Rinse thoroughly with tap water.
8. Rinse thoroughly with distilled water.
9. Dry at temperatures up to 150°F (66°C).
10. Flex the resultant brittle surface of the rubber using clean rubber or plastic gloves so that a finely cracked appearance is produced. Cyclizing hardens and slightly oxidizes the surface, permitting the necessary wetting of the surface by the adhesive. The finely cracked surface indicates that the rubber is ready for bonding. Light lacy lines on the surface indicate insufficient immersion time. Deep, course cracks and a thick crusty surface indicate excessive immersion.

9.5.3 Preparation of Butyl Rubber by the Chlorination Process

1. Scrub the surface in a nonionic detergent solution (2 to 3% by weight) at 120° to 140°F (49° to 60°C).
2. Rinse thoroughly in tap water.
3. Rinse thoroughly in distilled water.
4. Air dry.

5. Immerse for 1 to 3 minutes at room temperature in a freshly made solution consisting of (add ingredients in order listed):

Distilled water	97.0 pbw
Sodium hypochlorite (Clorox or Purex)	3.0 pbw
Hydrochloric acid (specific gravity 1.2)	0.3 pbw

6. Rinse thoroughly with distilled water.
7. Dry at temperature up to 150°F (66°C).

9.5.4 Priming

Prime with butyl rubber adhesive in an aliphatic solvent.

9.6 Chlorobutyl Rubber

This modified butyl rubber has greater vulcanization flexibility and enhanced cure compatibility than other general purpose rubbers.

9.6.1 Preparation of Chlorobutyl Rubber by Abrasive Treatment

1. Scrape the surface with a sharp blade to remove gross layers of wax, sulfur and other compounding ingredients which may have floated to the surface during processing.
2. Solvent wipe with toluene.
3. Uniformly abrade the surface with 80 to 120 grit abrasive paper. Machine sanding with a "jitterbug" oscillating sander is preferred over hand sanding because the machine sanding produces a more uniform surface with less effort. A buffing wheel may also be used.
4. Solvent wipe again with toluene to remove loose particles.

9.6.2 Preparation of Chlorobutyl Rubber by the Cyclization Process

1. Scrape the surface with a sharp blade to remove gross layers of wax, sulfur and other compounding ingredients which may have floated to the surface during processing.
2. Solvent wipe with toluene.
3. Immerse the rubber surface in concentrated sulfuric acid (specific gravity 1.84) for 5 to 45 minutes [see Step (10)].
4. Rinse thoroughly with tap water. Hot tap water is preferable.
5. Rinse thoroughly with distilled water.
6. Neutralize by immersion for 5 to 10 minutes in a 10 to 20% solution of ammonium hydroxide (sodium hydroxide may also be used).
7. Rinse thoroughly with tap water.
8. Rinse thoroughly with distilled water.
9. Dry at temperatures up to 150°F (66°C).
10. Flex the resultant brittle surface of the rubber using clean rubber or plastic gloves so that a finely cracked appearance is produced. Cyclizing hardens and slightly oxidizes the surface, permitting the necessary wetting of the surface by the adhesive. The finely cracked surface indicates that the rubber is ready for bonding. Light lacy lines on the

surface indicate insufficient immersion time. Deep, coarse cracks and a thick crusty surface indicate excessive immersion.

9.6.3 Preparation of Chlorobutyl Rubber by the Chlorination Process

1. Scrub the surface in a nonionic detergent solution (2 to 3% by wt) at 120° to 140°F (49° to 60°C).
2. Rinse thoroughly in tap water.
3. Rinse thoroughly in distilled water.
4. Air dry.
5. Immerse for 1 to 3 minutes at room temperature in a freshly made solution consisting of (add ingredients in order listed):

Distilled water	97.0 pbw
Sodium hypochlorite (clorox or Purex)	3.0 pbw
Hydrochloric acid (specific gravity 1.2)	0.3 pbw

6. Rinse thoroughly with distilled water.
7. Dry at a temperature up to 150°F (66°C).

9.6.4 Priming

Prime with butyl rubber adhesive in an aliphatic solvent.

9.7 Chlorosulfonated Polyethylene (Hypalon®)

This synthetic rubber, also known as chlorosulfonyl polyethylene, is characterized by ozone resistance, light stability, heat resistance, weathering, resistance to deterioration by corrosive chemicals and good oil resistance.

9.7.1 Preparation of Chlorosulfonated Rubber by Abrasive Treatment

1. Scrape the surface with a sharp blade to remove gross layers of wax, sulfur and other compounding ingredients which may have floated to the surface during processing.
2. Solvent wipe with acetone or MEK.
3. Uniformly abrade the surface with 80 to 120 grit abrasive paper. Machine sanding with a "jitterbug" oscillating sander is preferred over hand sanding because the machine sanding produces a more uniform surface with less effort. A buffing wheel may also be used.
4. Solvent wipe again with acetone or MEK to remove loose particles.

9.7.2 Priming

1. Wipe with toluene.
2. Force dry 1 hour at 140°F (60°C).
3. Priming with an adhesive primer SCOTCHCAST XR-5001 (3 M Co.) and BR-1009-8 (Cyanamid) have been reported satisfactory.

9.8 Nitrile Rubber

Nitrile rubbers are copolymers of butadiene and acrylonitrile and are frequently referred to as BUNA N. Properties vary with the acrylonitrile content. The nitrile rubbers exhibit a high degree of resistance to attack by oils at both normal and elevated temperatures. Carboxylated nitrile rubber exhibits many improved properties in comparison to the basic nitrile rubbers.

9.8.1 Preparation of Nitrile Rubber by Abrasive Treatment

1. Scrape the surface with a sharp blade to remove gross layers of wax, sulfur and other compounding ingredients which may have floated to the surface during processing.
2. Solvent wipe with methanol.
3. Uniformly abrade the surface with 80 to 120 grit abrasive paper. Machine sanding with a "jitterbug" oscillating sander is preferred over hand sanding because machine sanding produces a more uniform surface with less effort. A buffing wheel may also be used.
4. Solvent wipe again with methanol to remove loose particles.

9.8.2 Preparation of Nitrile Rubber by the Cyclization Process

1. Scrape the surface with a sharp blade to remove gross layers of wax, sulfur and other compounding ingredients which may have floated to the surface during processing.
2. Solvent wipe with methanol.
3. Immerse the rubber surface in concentrated sulfuric acid (specific gravity 1.84) for 10 to 15 minutes [see Step (10)].
4. Rinse thoroughly with tap water. Hot tap water is preferable.
5. Rinse thoroughly with distilled water.
6. Neutralize by immersion for 5 to 10 minutes in a 10 to 20% solution of ammonium hydroxide (sodium hydroxide may also be used).
7. Rinse thoroughly with tap water.
8. Rinse thoroughly with distilled water.
9. Dry at temperature up to 150°F (66°C).
10. Flex the resultant brittle surface of the rubber using clean rubber or plastic gloves so that a finely cracked appearance is produced. Cyclizing hardens and slightly oxidizes the surface, permitting the necessary wetting of the surface by the adhesive. The finely cracked surface indicates that the rubber is ready for bonding. Light lacy lines on the surface indicate insufficient immersion time. Deep, coarse cracks and a thick crusty surface indicate excessive immersion.

9.8.3 Preparation of Nitrile Rubber by the Chlorination Process

1. Scrub the surface in a nonionic detergent solution (2 to 3% by weight) at 120° to 140°F (49° to 60°C).
2. Rinse thoroughly in tap water.
3. Rinse thoroughly in distilled water.
4. Air dry.

5. Immerse for 1 to 3 minutes at room temperature in a freshly made solution consisting of (add ingredients in order listed):

Distilled water	97.0 pbw
Sodium hypochlorite (Clorox or Purex)	3.0 pbw
Hydrochloric acid (specific gravity 1.2)	0.3 pbw

6. Rinse thoroughly with distilled water.
7. Dry at a temperature up to 150°F (66°C).

9.8.4 Priming

1. Wipe with toluene.
2. Force-dry at 140°F (60°C).
3. Prime with adhesive primer.

9.9 Polyurethane Elastomers

Polyurethane elastomers have unique elastomeric properties which include exceptionally high abrasion resistance at moderate temperatures, excellent oil and solvent resistance, very high tear and tensile strength and high hardness with good mechanical strength. Both polyester and polyether types are available.

9.9.1 Preparation of Polyurethane Elastomers by Abrasive Treatment

1. Scrape the surface with a sharp blade to remove gross layers of wax, sulfur and other compounding ingredients which may have floated to the surface during processing.
2. Solvent wipe with methanol.
3. Uniformly abrade the surface with 80 to 120 grit abrasive paper. Machine sanding with a "jitterbug" oscillating sander is preferred over hand sanding because the machine sanding produces a more uniform surface with less effort. A buffing wheel may also be used.
4. Solvent wipe again with methanol to remove loose particles.

9.9.2 Priming

Primers are available for use with polyurethane elastomers and should be used in accordance with the manufacturer's instructions.

9.10 Synthetic Natural Rubber

This rubber is synthetic polyisoprene and it approximates the chemical composition of natural rubber. The synthetic polyisoprenes are lower in modulus and higher in elongation than the natural product.

9.10.1 Preparation of Synthetic Natural Rubber by Abrasive Treatment

1. Scrape the surface with a sharp blade to remove gross layers of wax, sulfur and other compounding ingredients which may have floated to the surface during processing.
2. Solvent wipe with methanol or isoprene.
3. Uniformly abrade the surface with 80 to 120 grit abrasive paper. Machine sanding with a "jitterbug" oscillating sander is preferred over hand sanding because the machine sanding produces a more uniform surface with less effort. A buffing wheel may also be used.
4. Solvent wipe again with methanol or isoprene to remove loose particles.

9.10.2 Preparation of Synthetic Natural Rubber by the Cyclization Process

1. Scrape the surface with a sharp blade to remove gross layers of wax, sulfur and other compounding ingredients which may have floated to the surface during processing.
2. Solvent wipe with methanol or isoprene.
3. Immerse the rubber surface in concentrated sulfuric acid (specific gravity 1.84) for 5 to 10 minutes [see Step (10)].
4. Rinse thoroughly with tap water. Hot tap water is preferable.
5. Rinse thoroughly with distilled water.
6. Neutralize by immersion for 5 to 10 minutes in a 10 to 20% solution of ammonium hydroxide (sodium hydroxide may also be used).
7. Rinse thoroughly with tap water.
8. Rinse thoroughly with distilled water.
9. Dry at temperatures up to 150°F (66°C).
10. Flex the resultant brittle surface of the rubber using clean rubber or plastic gloves so that a finely cracked appearance is produced. Cyclizing hardens and slightly oxidizes the surface, permitting the necessary wetting of the surface by the adhesive. The finely cracked surface indicates that the rubber is ready for bonding. Light lacy lines on the surface indicate insufficient immersion time. Deep, coarse cracks and a thick crusty surface indicate excessive immersion.

9.10.3 Preparation of Synthetic Natural Rubber by the Chlorination Process

1. Scrub the surface in a nonionic detergent solution (2 to 3% by weight) at 120° to 140°F (49° to 60°C).
2. Rinse thoroughly in tap water.
3. Rinse thoroughly in distilled water.
4. Air dry.
5. Immerse for 1 to 3 minutes at room temperature in a freshly made solution consisting of (add ingredients in order listed):

Distilled water	97.0 pbw
Sodium hypochlorite (Clorox or Purex)	3.0 pbw
Hydrochloric acid (specific gravity 1.2)	0.3 pbw

6. Rinse thoroughly with distilled water.
7. Dry at a temperature up to 150°F (66°C).

9.11 Styrene–Butadiene Rubber

This rubber, known as BUNA-S or SBR (GR-S old designation), is a copolymer of styrene and butadiene.

9.11.1 Preparation of Styrene–Butadiene Rubber by Abrasive Treatment

1. Scrape the surface with a sharp blade to remove gross layers of wax, sulfur and other compounding ingredients which may have floated to the surface during processing.
2. Solvent wipe with toluene.
3. Uniformly abrade the surface with 80 to 120 grit abrasive paper. Machine sanding with a "jitterbug" oscillating sander is preferred over hand sanding because the machine sanding produces a more uniform surface with less effort. A buffing wheel may also be used.
4. Solvent wipe again with toluene to remove loose particles.
5. Excessive exposure to toluene results in swollen rubber. A 20 minute drying time will restore the part to its original dimensions.

9.11.2 Preparation of Styrene–Butadiene Rubber by the Cyclization Process

1. Scrape the surface with a sharp blade to remove gross layers of wax, sulfur and other compounding ingredients which may have floated to the surface during processing.
2. Solvent wipe with toluene.
3. Immerse the rubber surface in concentrated sulfuric acid (specific gravity 1.84) for 10 to 15 minutes [see Step (10)].
4. Rinse thoroughly with tap water. Hot tap water is preferable.
5. Rinse thoroughly with distilled water.
6. Neutralize by immersion for 5 to 10 minutes in a 10 to 20% solution of ammonium hydroxide (sodium hydroxide may also be used).
7. Rinse thoroughly with tap water.
8. Rinse thoroughly with distilled water.
9. Dry at temperatures up to 150°F (66°C).
10. Flex the resultant brittle surface of the rubber using clean rubber or plastic gloves so that a finely cracked appearance is produced. Cyclizing hardens and slightly oxidizes the surface, permitting the necessary wetting of the surface by the adhesive. The finely cracked surface indicates that the rubber is ready for bonding. Light lacy lines on the surface indicate insufficient immersion time. Deep, coarse cracks and a thick crusty surface indicate excessive immersion.

9.11.3 Preparation of Styrene–Butadiene Rubber by the Chlorination Process

1. Scrub the surface in a nonionic detergent solution (2 to 3% by weight) at 120° to 140°F (49° to 60°C).
2. Rinse thoroughly in tap water.

3. Rinse thoroughly in distilled water.
4. Air dry.
5. Immerse for 1 to 3 minutes at room temperature in a freshly made solution consisting of (add ingredients in order listed):

Distilled water	97.0 pbw
Sodium hypochlorite (Clorox or Purex)	3.0 pbw
Hydrochloric acid (specific gravity 1.2)	0.3 pbw

6. Rinse thoroughly with distilled water.
7. Dry at a temperature up to 150°F (66°C).

9.11.4 Priming

A primer such as butadiene-styrene dissolved in an aliphatic solvent like methyl isobutyl ketone may be used.

9.12 Polybutadiene

This material is also known as butadiene rubber.

9.12.1 Preparation of Polybutadiene by Abrasive Treatment

1. Scrape the surface with a sharp blade to remove gross layers of wax, sulfur and other compounding ingredients which may have floated to the surface during processing.
2. Solvent wipe with methanol.
3. Uniformly abrade the surface with 80 to 120 grit abrasive paper. Machine sanding with a "jitterbug" oscillating sander is preferred over hand sanding because the machine sanding produces a more uniform surface with less effort. A buffing wheel may also be used.
4. Solvent wipe again with methanol to remove loose particles.

9.12.2 Preparation of Polybutadiene by the Cyclization Process

1. Scrape the surface with a sharp blade to remove gross layers of wax, sulfur and other compounding ingredients which may have floated to the surface during processing.
2. Solvent wipe with methanol.
3. Immerse the rubber surface in concentrated sulfuric acid (specific gravity 1.84) for 10 to 15 minutes [see Step (10)].
4. Rinse thoroughly with tap water. Hot tap water is preferable.
5. Rinse thoroughly with distilled water.
6. Neutralize by immersion for 5 to 10 minutes in a 10 to 20% solution of ammonium hydroxide (sodium hydroxide may also be used).
7. Rinse thoroughly with tap water.
8. Rinse thoroughly with distilled water.
9. Dry at temperatures up to 150°F (66°C).
10. Flex the resultant brittle surface of the rubber using clean rubber or plastic gloves so that a finely cracked appearance is produced. Cyclizing hardens and slightly oxidizes the surface, permitting the necessary wetting of the surface by the adhesive. The finely

cracked surface indicates that the rubber is ready for bonding. Light lacy lines on the surface indicate insufficient immersion time. Deep, coarse cracks and a thick crusty surface indicate excessive immersion.

9.12.3 Preparation of Polybutadiene by the Chlorination Process

1. Scrub the surface in a nonionic detergent solution (2 to 3% by weight) at 120° to 140°F (49° to 60°C).
2. Rinse thoroughly in tap water.
3. Rinse thoroughly in distilled water.
4. Air dry.
5. Immerse for 1 to 3 minutes at room temperature in a freshly made solution consisting of (add ingredients in order listed):

Distilled water	97.0 pbw
Sodium hypochlorite (Clorox or Purex)	3.0 pbw
Hydrochloric acid (specific gravity 1.2)	0.3 pbw

6. Rinse thoroughly with distilled water.
7. Dry at a temperature up to 150°F (66°C).

9.12.4 Preparation of Polybutadiene by Solvent Wiping

A simple solvent wipe with methanol may be used.

9.13 Fluorosilicone Elastomers

Fluorosilicone rubbers retain most of the useful qualities of the silicone rubbers and improved resistance to many solvents, with the exception of the ketones and phosphate esters. They have good low-temperature resistance.

9.13.1 Preparation of Fluorosilicone Elastomers by Solvent Cleaning

Wipe with methyl, ethyl or isopropyl alcohol or toluene.

9.14 Epichlorohydrin Rubbers

These rubbers exhibit exceptional resistance to aliphatic hydrocarbons, ozone, weathering, gas permeability, compression set, impact and tear. They have good tensile strength, resilience, and resistance to abrasion, water, acids and alkalis.

9.14.1 Preparation of Epichlorohydrin Elastomers by Solvent Cleaning

Solvent wipe with acetone, alcohols, or aromatics such as toluene. Do not use methyl ethyl ketone (MEK).

9.15 Polysulfide Rubber

9.15.1 Preparation of Polysulfide Rubber by Abrasive Treatment

1. Scrape the surface with a sharp blade to remove gross layers of wax, sulfur and other compounding ingredients which may have floated to the surface during processing.
2. Solvent wipe with methanol.
3. Uniformly abrade the surface with 80 to 120 grit abrasive paper. Machine sanding with a "jitterbug" oscillating sander is preferred over hand sanding because the machine sanding produces a more uniform surface with less effort. A buffing wheel may also be used.
4. Solvent wipe again with methanol to remove loose particles.

9.15.2 Preparation of Polysulfide Rubber by the Chlorination Process

1. Degrease in methanol.
2. Immerse overnight in strong chlorine water.
3. Wash with water.
4. Dry.

9.16 Polypropylene Oxide Rubber

This material performs similarly to natural rubber, but has the added advantages of good resistance to aging at high temperature, good ozone resistance and moderate resistance to loss of properties in contact with fuels and some solvents.

9.16.1 Preparation of Polypropylene Oxide Rubber by Solvent Cleaning

Wipe the surface with TCE, xylene, toluene, or other appropriate solvent.

9.17 Fluorocarbon Elastomers

These elastomers have excellent resistance to ozone, oxidation, weathering, heat, aliphatic and aromatic hydrocarbons and alcohols. They are highly impermeable to gases, have good strength, electical resistivity, and resistance to abrasion, water, acids and halogenated hydrocarbons.

9.17.1 Preparation of Fluorocarbon Elastomers by the Sodium-Naphthalene Complex Process

1. Wipe with acetone.
2. Treat with sodium-naphthalene complex for 15 minutes at room temperature. Sodium-naphthalene complex solutions are commercially available (see ASTM D2093). Sodium-naphthalene complex solutions should be kept in tightly stoppered glass

containers to exclude air and moisture. Extreme care should be taken in handling these solutions, since they are hazardous, due to the explosive property of the sodium.

3. Remove from solution with metal tongs.
4. Wash with acetone to remove organic material.
5. Wash with distilled water to remove the last traces of metallic salts from the treated surface.
6. Dry in an air-circulating oven at 99.5°F (37.3°C) for about 1 hour.

9.18 Polyacrylate or Polyacrylic Rubber

These rubbers exhibit high resistance to heat, ozone, oxidation, weathering, aliphatic hydrocarbons and sulfur-bearing oils. They also have good resilience and gas impermeability.

9.18.1 Preparation of Polyacrylate or Polyacrylic Rubber by Abrasive Treatment

1. Wipe or spray with or immerse in 1,1,1-trichloroethane, FREON TMC or M-17, acetone, MEK, toluene, methyl, ethyl or isopropyl alcohol.
2. Abrade lightly and uniformly with 180 to 320 grit abrasive paper.
3. Repeat Step (1).

Reference

[1] R.T. McIntyre, et al., Effect of Varying Processing Parameters in the Fabrication of Adhesive Bonded Structures, Part VI, Production Methods, Picatinny Arsenal Technical Report 4162, Picatinny Arsenal, Dover, NJ, February 1971.

Index

Note: Page numbers followed by "*f*" and "*t*" refer to figures and tables, respectively.

Printed and bound by CPI Group (UK) Ltd, Croydon, CR0 4YY

08/05/2025

01864838-0004